Geographic Information Systems and Science

Edited by Jorge Rocha and Patrícia Abrantes

Published in London, United Kingdom

IntechOpen

Supporting open minds since 2005

Geographic Information Systems and Science
http://dx.doi.org/10.5772/intechopen.75243
Edited by Jorge Rocha and Patrícia Abrantes

Contributors
Wenjie Huang, Mengzhi Ling, Salem Issa, Nazmi Saleous, Hamdan Omar, Muhamad Afizzul Misman, Samsudin Musa, Lien Kwei Chien, Jing-Ping Wu, Wen-Chien Tseng, Ming Lu, Duanshun Li, Rod Wales, N N Ramaprasad, Priya Narayanan, Kayoko Yamamoto, Koya Tsukahara, Junghoon Ki, Jorge Rocha, Patricia Abrantes, Cláudia Morais Viana

First published in London, United Kingdom, 2019 by IntechOpen
IntechOpen is the global imprint of INTECHOPEN LIMITED, registered in England and Wales, registration number: 11086078, 7th floor, 10 Lower Thames Street, London, EC3R 6AF, United Kingdom
Printed in Croatia

British Library Cataloguing-in-Publication Data
A catalogue record for this book is available from the British Library

Additional hard and PDF copies can be obtained from orders@intechopen.com

Geographic Information Systems and Science
Edited by Jorge Rocha and Patrícia Abrantes
p. cm.
Print ISBN 978-1-83962-233-5
Online ISBN 978-1-83962-234-2
eBook (PDF) ISBN 978-1-83962-235-9

For EU product safety concerns:
IN TECH d.o.o., Prolaz Marije Krucifikse Kozulić 3, 51000 Rijeka, Croatia, info@intechopen.com or visit our website at intechopen.com.

Meet the editors

Jorge Rocha is a geographer. He has an MSc in Geographic Information Systems (2003) and Spatial Planning (2013), and a PhD in Geographic Information Science (2012). Currently, he is an assistant professor at the Institute of Geography and Spatial Planning and a member of the Modelling, Urban and Regional Planning, and Environmental Hazard and Risk Assessment and Management research groups of the Centre of Geographical Studies, all at the University of Lisbon. His field of expertise is geosimulation and geocomputation, involving artificial neural networks, graphs theory, cellular automata, and multiagent systems. Jorge's works are quite diverse, focusing mainly, but not exclusively, on urban morphology, remote sensing, epidemiology, health geography, smart cities, and big data.

Patrícia Abrantes is an assistant professor in the Institute of Geography and Spatial Planning, Universidade de Lisboa. In 2007, she completed her PhD in Geography, specializing in spatial planning in the University Nova of Lisbon and in Paris IV (Sorbonne). In the last few years, she has developed several research projects, including DAUME (sustainability of urban agriculture in the Mediterranean area) and AgriMet-MOD (Metropolitan Agriculture spatial MODeling: contributing to the construction of an analysis model to support spatial planning and land use decision-making).
Her research interests are centered on the application of geographic technologies to spatial planning, in particular in metropolitan spaces. Her subject areas include land use and urban agriculture, agriculture ecosystem services, spatial analysis, geographic information systems (GIS), and participatory GIS.

Contents

Preface

Geographic information systems (GIS) were seen as a panacea for quite a long time. The misfortune is that they have become more like service-oriented architecture solutions. Indeed, we previously stored our files on either a PC or a server. However, currently files are no longer allocated; instead, we use several network services such as a Web Map Service (WMS) or Web Feature Service (WFS), which distribute network services. We have definitely moved from a web supported by documents to a web supported by databases, crowdsourced data (e.g., volunteered geographic information—VGI), and social networks, twisted by community behavior, giving access to both collaboration tools and environments, and allowing location analytics.

We have reached a point where anybody may gather spatial-referenced data using, for instance, a mobile device, create some reasonable maps, and put them online. VGI is effectively replacing official data sources, letting us perform a more complex and dynamic analysis than the one allowed by traditional census data. This we can call neogeography, a new kind of geography accessible to everyone. Neogeography offers new life to maps, thriving in a society where each individual is a potential cartographer. Now maps are able to display individual perceptions because they are centered on the mapmaker, representing reality from a bottom-up perspective rather than from a bottom-down one. This new generation of maps can be created in real time and be tailor-made, representing detailed singularities never seen before, e.g., graffiti on a wall or birds in a tree.

It is clear that data-driven geography is re-emerging due to substantial spatial data flow from people and sensors. Geography has shifted from a data-scarce to a data-rich environment. This—big data—revolution is not about data volume; instead, it is about data variety and at what velocity we can collect and store it. Big data has an enormous potential to feed both spatial analysis and geographic knowledge-discovery but at the same time raise friction among idiographic and nomothetic methodologies. Nonetheless, the belief that location matters is inherent to geography and functions as a robust incentive to develop sophisticated procedures on spatial statistics, time-geography, and geographic information science (GISc).

GISc can be understood as a subdivision of information science that deals with geographic data, or as a set of vital interrogations upraised by geographical information and the technologies used to gather, handle, and communicate it, i.e., information and communication technologies (ICTs). Geographic data can disclose fascinating patterns that, in specific cases, point toward causal mechanisms.

GIS have evolved from a research project to a very profitable industry. As new, quicker, more powerful, and less expensive technologies become accessible, this tendency will carry on. The new generations of mobile phones have capabilities to perform GIS tasks, augment the reality recognized by the handler by getting into databases, and have the capacity to capture and upload photos. Furthermore, the VGI invisible economy greatly increases the value of the visible economy, because people are interested in becoming neogeographers. GIS can be extremely seductive, tempting people to experience an intrinsic love for maps.

Certain key issues are sure to endure in the nearby future. The spatial heterogeneity principle makes sure that most probably local and universal will be struggling with each other, claiming to harmonize standards and improve software and data interoperability. Data interoperability and Web services are ever more crucial, forcing, to some extent, the replacement of the GIS concept with spatial data infrastructures. The spatial dependence standard guarantees that efforts to apply orthodox techniques of inferential statistics to geographic phenomena will always be troubled and the spatial heterogeneity principle will always contradict nomothetic knowledge.

GIS are becoming network-oriented services. In these services, only organized internal resources become worldwide network-shared services. However, network services and smartphones, to name just a few, will soon claim interoperability, openness, and flexibility. Several GIS functions will persist for a long time. Network intelligence is being added, e.g., WMS, WFS, and CityGML; a completely new revolution is on the move. So, obviously, GISc will carry on, but rapidly change and evolve as well. The main GISc challenge is to find useful and efficient ways to catch and map the complexity of geographic systems in the limited digital binary space of a computer. In addition, we face the challenge of depicting the remainder and evaluating its influence on GIS operation results. This generates a question of critical spatial thinking, i.e., the critical process that any user of these technologies must have.

Undeniably, recent developments in ICTs have increased the multimedia narratives about the geographical representation of places, with significant repercussions on geographic research directions. Geographers incorporating big data with current scientific paradigms have changed and supported the study of geographic systems and, in between, settled for new concepts of space. This is a break for new research agendas in both qualitative and quantitative contexts. Big data and data analytics improve our understanding of the use of space, evaluating physical (fixed) and digital (fluid) spaces, and while both overlap and coexist, each one is shaped by the other and its users.

So, one may conclude that over the last few years, GISc has established itself as a collaborative information-processing system that is increasing in popularity. Yet, this interdisciplinary field is still somewhat misunderstood. This book talks about some of the GISc domains encompassing students, researchers, and common users. Chapters focus on important aspects of GISc, keeping in mind the processing capability of GIS along with the mathematics and formulae involved in arriving at each solution. The book has nine chapters divided into five sections. The first section is more general and focuses on what GISc is and its relation to GIS and Geography, the second is on location analytics and modeling, the third on remote sensing data analysis, the fourth on big data and augmented reality, and finally, the fifth is focused on VGI.

Jorge Rocha and Patrícia Abrantes
Institute of Geography and Spatial Planning,
University of Lisbon,
Lisbon, Portugal

Section 1

Geographic Information Systems and Science

Chapter 1

Introductory Chapter: Geographic Information Systems and Science

Cláudia M. Viana, Patrícia Abrantes and Jorge Rocha

1. Introduction

Information science can be defined as the science that investigates the properties and behavior of information, the forces that govern its flow, and the means of processing it. The process includes the origin, dissemination, collection, organization, storage, retrieval, interpretation, and use of information. Although this view is not consensual, one can state that there are several information sciences and they can be separated into two groups. The first group (e.g., librarianship, journalism, and communication) mainly studies the transfer of information and the second group (e.g., philosophy, sociology, and cognitive sciences, in general) focuses preferably on certain aspects of information transfer.

Nonetheless, it is not easy to think of a science in which at least one of its study objects does not consider the geographic space. However, it is evident that some areas today have appropriated the geographical information more than others have. Assuming that data (geographic or not) is a construction of the knowledge within a certain political, economic, and spatiotemporal context, then the question is how it is possible to set apart the geographical, or in a more broad sense the spatial, context of these data?

Spatial knowledge has long been critically important both in the development of human activities and in the understanding of how humankind interacts with space. The possibility of knowing the geographical location of a phenomenon, to establish relationships with other elements of the territory, to identify spatial patterns, or to make decisions based on their geographical examples of how spatial information and knowledge have over the years a preponderant role in social, political, and economic development.

The geographic information of the world is in a period of exponential development, sustained by the availability of new sensors, imageries, open georeferenced data and social media, and moving devices, allowing to produce new knowledge about that world and empirical geographical phenomena [1, 2]. Almost all the phenomena we face daily assume a territorial expression. Represent, analyze, and know the spatial dimension associated with these phenomena is one of the main challenges faced by the Geographic Information Systems (GIS).

The continuous collection of geographic data of holistic nature and the analysis of its place in territory, i.e., location analytics, allows a more straightforward science [3] that can influence the creation and the use of geographic knowledge. This holistic point of view allows researchers to put questions that before were impossible [4]. Multidisciplinary approaches take along important knowledge about the adoption of system approach to information management.

Just as the information science can be understood as a group, geographical science is a set of sciences interested on physical and/or human geographic processes.

IntechOpen

Although this definition is consistent with the broad definition of geography, the term should not only be circumscribed to the current school of Geography, but also involving areas such as meteorology, geology, environment, epidemiology, geomarketing, and many others that study the phenomenon's geographical dimensions. In fact, today, GIS brings together several areas of knowledge. These areas of science influence each other, creating conceptual and technical interfaces such as multidisciplinarity.

Michael Goodchild first coined the designation "Geographic Information Science" in 1992, defining it as the persistent research, consistent with methodical values, of the type and characteristics of the data [5]. This science operates through the techniques, methods, and approaches associated with GIS and seeks to redefine the geographic concepts and their use in the context of GIS. Thus, it becomes clear that the term Geographic Information Science (GIS) is also multidisciplinary.

2. Is GIS a science?

Is GIS a science? In 1992, Goodchild [5] was asked to write a paper defining what could be Geographic Information Science. At that time, GIS became more generalized in enterprises and in research, becoming also popular in training programs, so GIS community began to view it as more than just a tool or system as they started to research GIS itself.

Goodchild paper gave a major contribution in coining the term GIscience or GISc; however, the author did not propose a definition for this science. Instead, he addressed the uniqueness of geographical data in terms of its properties and the need of particular methodologies to deal with it [5], and he presented eight major contents that this new science should address, namely: information gathering and quantification; data acquisition; geo(spatial) statistics and other location analysis tools; spatiotemporal modeling and spatial concepts; spatial data infrastructures, algorithms and procedures; visualization; and decision-making, administrative and moral concerns.

Almost 30 years passed since Goodchild wrote this paper entitled Geographic Information Science and discussing the science in GIS. Even though 30 years have passed, it is curious to note that:

1) The contents are up-to-date. For instance, in data collection and measurement, we still discuss challenges regarding generalization and abstraction; data capture also possess challenges now much more related to the massive production of data partly due to the web and Internet of Things (IoT) developments, and the quality of that data in a time where almost everyone can produce geographic data. In addition, in display domain, we passed from problems related to 2.5 and 3D to 4D, 5D, virtual reality, and immersion. On the other hand, in relation to institutional, managerial, and ethical issues, the new improvements in the web raised more challenges related to free data software and interoperability, and to volunteer geographic information (VGI) and the production and use of data, or to geographic information (GI) access disparities and data privacy problems.

2) Whether GI is a science or it is there to support other sciences continues to be a long-standing debate. Why is it so? Because it lacks its own law, it applies (adapt) other science algorithms and theories, so it cannot exist independently from the other sciences, and it is extremely dynamic with new branches being added frequently and with it added are fuzzy boundaries with other sciences [6, 7].

What makes GI a science? Reitsma [6] presented an interesting argument on this issue. He supports that GI is a science because it has a distinctive object of study and the (geographic) representation of the world, as Goodchild, in 1992, once referred to when he argued about the uniqueness of spatial (geographic) data. Therefore, how to transform spatial data into geographic thinking or knowledge is the major core of this science [5].

Is there a definition for GISc? All said, we might well think that there is no definition for GISc. In fact, defining GISc is not consensual, just as GIS for long time. Nevertheless, there are some attempts. Rapper's definition is perhaps the most well known as he defines GISc as "a perfect multidiscipline with a core of theory, data, and software engineering work and a periphery of engagement with related disciplines"[8]. While Reitsma [6] suggests that it is the study of how geographic information is formed, collected, managed, analyzed, and visualized to represent the world. Others added that GISc is also responsible to analyze the impacts of GIS in society and vice versa [8, 9].

3. GIScience

Geographic Information Science (GIS) operates through the techniques, methods, and approaches associated with GIS and seeks to redefine the geographic concepts and their use in the context of GIS. Thus, it becomes clear that the term GISc is also multidisciplinary.

Here, we begin to propose some concepts more comprehensive about geography, which treats "Geo" not only as a discipline, but also as a means to an end. For this, first, we first go to the conceptualization of science as a body of systematized knowledge acquired through observation, identification, research, and explanation of certain categories of phenomena and facts and formulated methodically and rationally.

Going a little further, we came to this definition of Geographic Information Science, by Mike Goodchild, a researcher at the University of Santa Barbara who advised former President of the United States of America, Al Gore, on the movie "Inconvenient Truth." "Information Science can be defined as the systematic study, according to scientific principles, of the nature and properties of information. Geographic Information Science is the subset of information science that is about geographic information" [5].

That is, does it make sense to talk about Geographic Information Science? Or is it ok using a shorter term Geoscience? Indeed, the concept of Digital Earth was introduced by former United States of America vice president Al Gore in 1998 [10]. Gore compared it to a digital world that mirrors the reality [11]. This concept becomes a reality thanks in large part to virtual globes (like Google Earth) that ease the massification of both the search and the sharing of information [12].

It is projected that Digital Earth should be a space for sharing of global information for development between regions and generations [13], and it is considered that the integration of this parallel world in the daily life of humanity will already be put in place in 2020 [12]. This context is perceived as an added value for geography. In addition to reinforcing its importance as a science of innovation, it still benefits from a new dimension of space, which leads to virtual geography, supporting the development of a geography with new contours, supported by the proliferation of new digital technologies [14]. The contents that are part of the GISc are:

- Data collection—all forms, from total stations, Global Navigation Satellite System (GNSS) receivers, satellites and drones, smartphones, devices of IoT and to the users themselves, who are also "walking sensors."

- Storage platforms and data management—here, we talk about anything that can store some kind of data or information, which can be in local servers, in the cloud, or in any other place.

- Data modeling, algorithms, and processes—the different data formats—vector, raster, point clouds, tables, and so on—how are they modeled, converted, and used.

- Data visualization—a few decades ago printed maps were the visualization standard but the emergence of computation led cartography to the screens. Today the geographic information is in the palm of our hand (e.g., smartphones), but also in new platforms such as virtual and augmented reality glasses, 3D printers, etc.

- Data analytics tools—not long ago, the power of layering or mapping the best route between multiple points was restricted to "semigods who inhabited the geoisland," but today anyone has the power of GIS in two clicks away (neogeography [15]).

- Institutional aspects—this is the item that makes it clear that GISc is much more than geotechnology because institutional issues involve politics, conflict of interests, cultural differences, disputes over ownership of data, and many others.

- Data sharing—finally, the ways of sharing information have changed a lot and will advance even more rapidly in the near future. Nowadays, novelties cross the world in a few seconds through social networks. This leads us towards a set of paradigm shifts about who owns the data, what is the reach of the information, what will be the impact of it, and about who will be the end users.

GISc must focus in the essential concerns that come from GI. Further expressions have more or less a similar connotation, e.g., spatial information science, geomatics, geoinformatics, and geocomputation. They all will advocate a scientific attitude regarding the core subjects upstretched by GIS applications and associated technologies. They all also have diverse backgrounds and highlight unalike means of facing geographic problems.

Finally, we arrive at the pyramid (**Figure 1**) that clearly shows the three dimensions of Geographic Information Science [16], each one of equal importance. One should note that of its three vertices, only one is related to technology, which is the dimension related to computing (although originally it was written computer, we can substitute here for any device with capacity to collect, store, analyze, and share geographic information). The other two pillars of the GISc relate to people and society in general. These are the vertices of a triangle, and GISc places itself in the

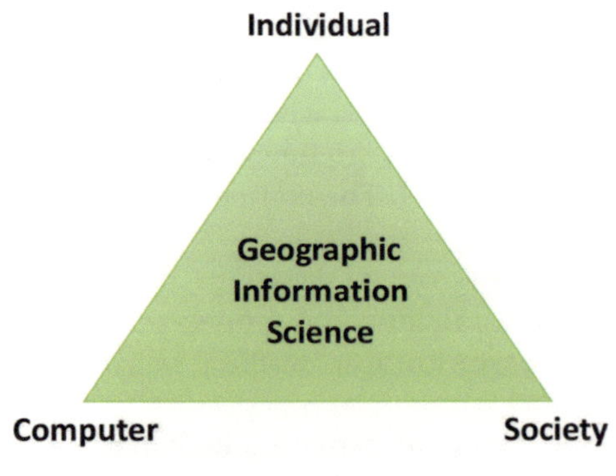

Figure 1.
Dimensions of geographic information science [16].

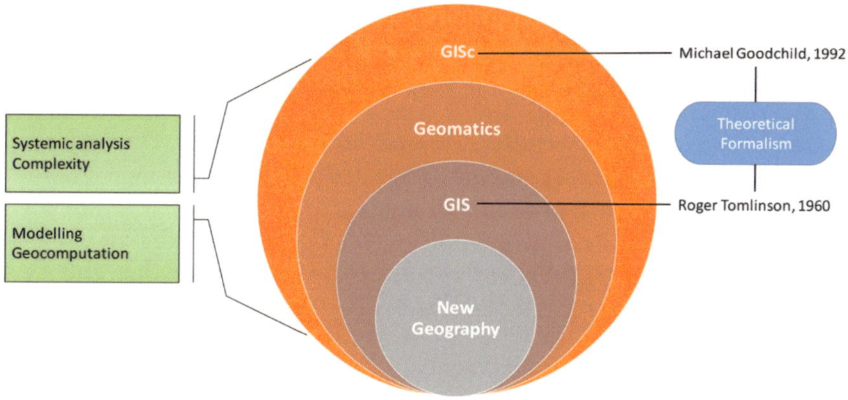

Figure 2.
From geography to GIS.

center. The several expressions used to refer to GISc may serve to fill this triangle: That is, terms like geotechnology, geomatics, and geoprocessing are valid for some approaches, but GISc is much more than that.

The field of GISc has only been recently evidenced as a scientific domain with autonomy, and not simply as an instrument or just an occasional merge of knowledge from other areas. In its fundamental component, it includes matters of cartography, positioning, information systems, computer graphics, and more. The exploitation of information can also make use of knowledge in the field of statistics, operational research, expert systems, and decision theory. The incorporation of concepts and the specific problems of the numerous fields of application brought together all these general problems of geographic modeling, adding to them theories and techniques in fields as diverse as physics, geography, planning, health, demography, and among many others.

One of the major challenges of GISc is the development of techniques and abstractions that are capable of adequately representing dynamic phenomena. Indeed, the GISc corresponds to a set of scientific knowledge and methods fundamental to the development and validation of spatial theories, but they do not advocate GIS as an isolated entity, i.e., a science per se (**Figure 2**). This approach agrees with the definition of GIS as a tool but considers it reductive, since it excludes concepts; fundamentals; methodologies; rules; and methods of analysis, measurement, evaluation, and decision support of spatial models.

Thus, there are significant issues where GISc is all about using a GIS environment to remake, restructure, and solve preexisting research problems, many of them much older than GIS itself. Indeed, it is much easier to label GISc than to define it [17]. Therefore, scientists usually adopt a defensive position when it comes to positioning GISc in relation to other sciences [6, 18].

Nevertheless, they all agree that basic and applied science should have its reflection in society [19]. Hence, one may determine that GISc is extremely pertinent to society. This is more difficult to demonstrate them compared to the broader term of Geographic Information Technologies (GIT), which many times is shown to be capable of influencing the society and vice-versa [20, 21].

4. Geography and GIScience

At present, we are facing a paradoxical situation. On one hand, there is the emergence of neogeography and the proliferation of user-generated geographic content and more precisely volunteered geographic information (VGI). On the other hand, we are witnessing a new incursion of physics into social sciences, with a growing motivation for the use of physical models in the analysis of social systems, e.g., cities [22–25] and social networks [26].

We are turning our scope to the traditional latent tension between macrogeography, centered in general principles (law-seeking), and microgeography (description-seeking), which proliferated in most of the twentieth century. The former advocates a nomothetic geographic knowledge and the later an idiographic one [27]. This is not the first time, and surely not the last, we see this, and we have differential modeling to prove it [28]. GISc holds both, the first in algorithms and methods and the second in data. One can easily draw a connection between physics and GISc, passing through spatial analysis [29]. This is the first insight that we should retain, and this is particularly true for human geographers, in which spatial context does matter.

Despite all the efforts, human geography has denied the importance of physical principle-based models, which makes it clear that the interactions between low-level system components can produce new characteristics that proved to be unpredictable, even if we fully understand the central laws that rule the system, i.e., emergency. These characteristics are obvious human systems of (auto-)organized complexity and sophisticated feedback loops (positive and negative) driving to emergent behaviors and counterintuitive effects, e.g., cities [30]. Here, spatial context is fundamental, because it influences the local pattern of interactions between system components and consequently the system dynamics that emerge from individual behaviors [31] in a phenomenon where one could designate aggregate complexity.

The use of models in any branch of geography research has proven to be more efficient than the traditional techniques of data analysis. Stating so, one do not intend to appeal to the rejection of any other technique that has demonstrated its usefulness, neither do we expect that geographers change their research objectives. The use of models at all levels is presently a different question. Models are so effective and the functional explanation of a system is much more effective than a set of disconnected facts that the use of models will prove to be of the most importance in the long run.

To advocate the use of models is not necessarily to insist on the study of general geography, spatial distribution, or the purpose of formulating general laws. In the long term, the most significant result of developing geographic models will be the establishment of genuine principles, distinct from superficial generalizations. These models are supported by a stronger basis and are enhanced with values that emerge at higher levels. For instance, considering spatial interaction models, or one of its outmost representatives: gravity model, it was first drawn as an analogy to the third Newton's gravity Law, then strengthened through entropy maximization [32, 33], and finally turned spatially explicit. The outcome was a refined panoply of spatial interaction models, including the ones focused on origin-destination spatial context [34].

This is where geography is different from natural sciences. It is where the exceptions, the discrepancies, and the uniqueness cannot be ignored because of the fact that the Earth's surface is not an isotropic space. One should not end up studying the discrepancies between the normative model and the real case, instead one should insist in finishing the study of the particular cases, of the normative element, of the special element, and of all of them as one. Nowadays, we have the knowledge and necessary techniques to study what appear to be parts of a more general system. Geography will just achieve internal consistency by being able to abstract himself from reality.

The choice is not only between human and physical, regional and general geography, nor is it only between regional differentiation, landscape evolution, human ecology, and spatial distribution and the recognition of spatial patterns, but may also be between objective academic studies and practical spatial planning. It may be between pointing out objectives to build coherent theories to search for order in a complex world and trying to understand those parts. In both, models are inevitable, stimulating, and economic.

One may enquire whether we are looking for truth or utility: a pertinent question currently in geography. Many simulation models used for prediction and planning are black box models and may be useful in a short-term practical application even presenting a false assumption of the systems nature. We can name the abductive reasoning with this type of logical inference, which starts with data and finishes by drawing a hypothesis that best fits that data. This kind of tentative knowledge (might be true) is more fragile than induction (what is true) and deduction (what must be true) knowledge. Therefore, it should be validated and sustained by far-reaching theories [35].

One factor that has gained importance and helped to reinforce the role of quantitative geography (and to dispel fears) about model implementation is the growing mathematical component, sometimes leading to the designation of mathematical geography. At present, we recognize geomatics, i.e., mathematical geography, as a branch of geography. Thus, the concept establishes a parallelism between geography and algorithms. Hence, it can be said that quantitative geography, well seconded by the diffusion and rapid growth of the computational technology, has unfolded toward what would be the logical evolution, the appearance of the GIS. This branch has expanded too many other disciplines that use GIS and remote sensing (**Figure 3**) [36]. **Figure 3** expresses advances pushing geographic science beyond cartography into a far more multipurpose and dominant vision of "maps" that has its reflection on many sciences and/or technologies.

If GIS traditionally has a singular connexion to geography, as it has to other sciences that deal with georeferenced data, e.g., engineering and landscape architecture, GISc,

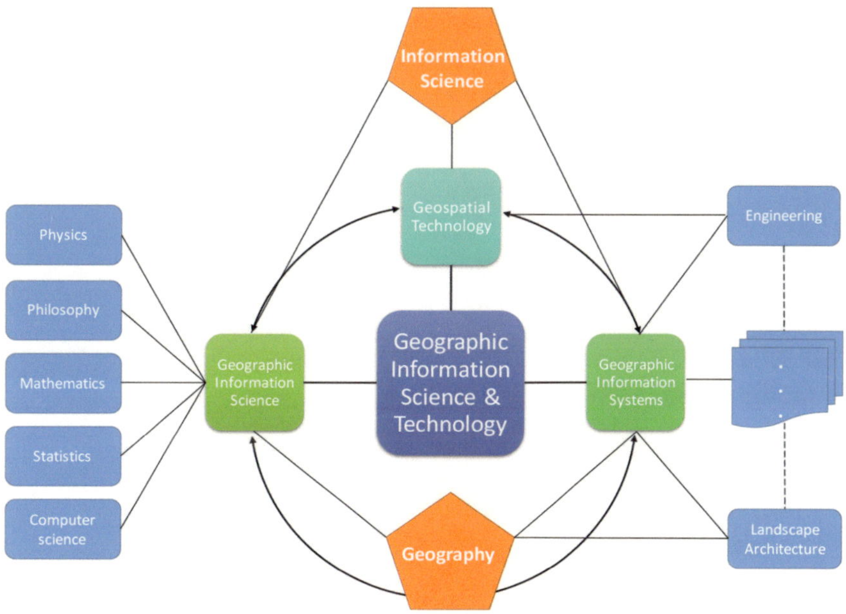

Figure 3.
From geography to GIS [36].

being an original set of concepts and expertise with an extensive applicability, has captured the attention of several traditional sciences, e.g., physics, philosophy, mathematics, etc. (**Figure 3**). As follows, GISc is many times labeled as interdisciplinary, multidisciplinary, transdisciplinary, and multiparadigmatic. Despite having different meanings, these definitions usually are used as synonymous. But, why this debate of science, subscience, or multidisciplinary field remembering the outdated—or perhaps not—discussion about if GIS and quantitative analysis are geography?

Some researchers still say that GISc is just a response to GIS technology and not a state-of-the-art science, giving the example of critical GIS, today fully integrated in GISc but initially the focus of pronounced disagreement [37]. This bifurcated propensity is clear in two polemic and completely distinct understandings [37]: one defending a mathematical and formal perspective of GISc, mainly focusing on quantitative and computational aspects [38], and other antiessentialist and anti-interpretative [39]. The latest one, being a deflationary method, allows joining a role of essential ideologies. However, taken to the extreme, it can lead to a totally amorphous field of research.

5. Insights into the future

According to Haklay [37], GISc relies upon an inductive approach (**Figure 4**). Contrarily to deductive approach that start with a theory that can either be corroborated or not, depending on the results of the validation tests, i.e., observations, carried out, inductive approach comes to theory based on observations, trying to detect patterns that can lead to the formulation of the former.

Remler and Van Ryzin [40] propose a clear division between observational studies, natural experiments, and weak/strong quasi-experiments (**Figure 5**). This distinction is made in function of the sorting method, if it has some degree of control (latter two) or if the researcher has no control over it (former two). Observation, i.e., evidence based approach, is supported by longitudinal and experimental procedures that reinforce the randomness prerogatives inherit from cross-sectional methods [41]. The new methods for data gathering and the availability of open big data enable this kind of approach for studying human-physical, or simply human, complex systems [42].

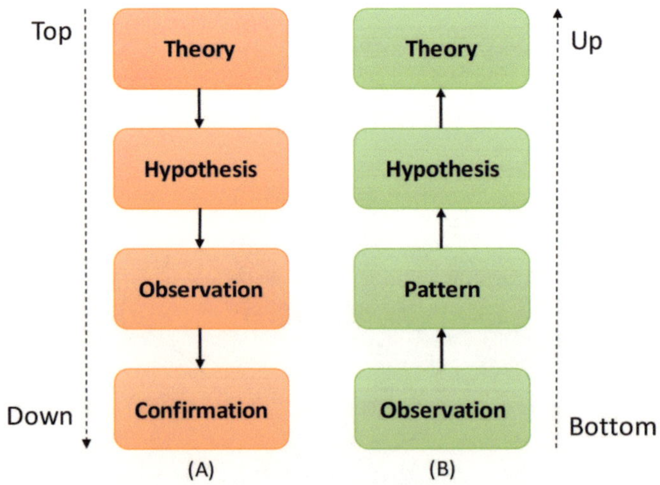

Figure 4.
Deductive (A) and inductive (B) workflows.

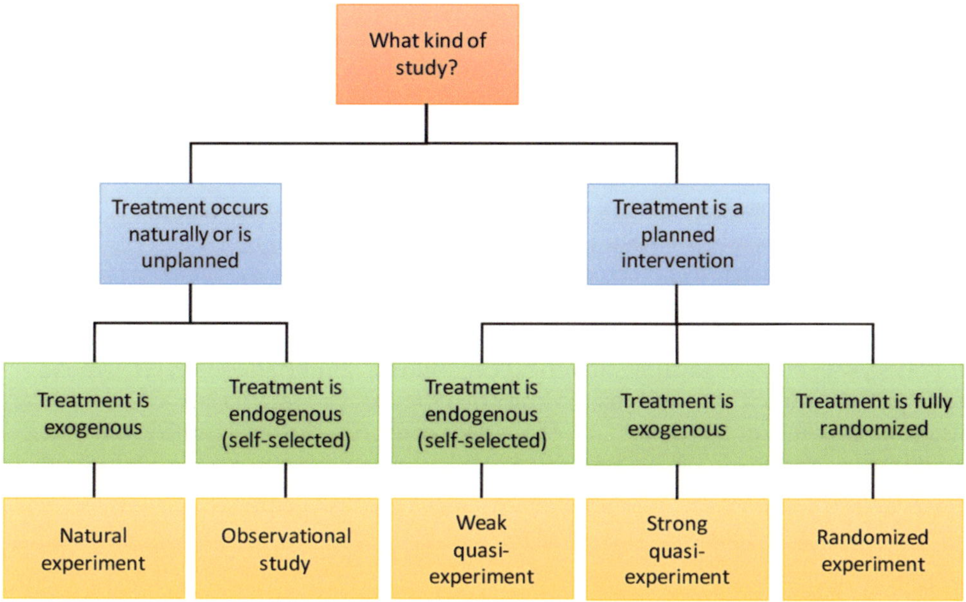

Figure 5.
Different approaches of experimental research [40].

The colossal and distinct data dissemination through a multitude of worldwide connected databases lead to an increasing expansion of complex data availability, built from different sources, and to its interaction with the existing procedures and behaviors [43, 44].

The increasing growth of popularity in social media platforms, e.g., Facebook, Twitter, Instagram, or Flickr, etc., resulted in the availability of a large amount of VGI. Goodchild [45] defines VGI as geographic data produced by users usually with the backing of Web 2.0 capabilities [46]. Indeed, society changed from a web supported by documents to a web supported by databases, crowdsourced data and social networks [47], bent by social behavior [48] and giving access to both collaboration tools and environments, allowing analytical visualizations [49].

The Web 2.0 and the omnipresence of the data radically changed not only the technical support of GIS but also of GISc, extending the potentials of people participation in administration and planning processes [50]. This holistic point of view allowed researchers to study social phenomenon using the digital traces and social interactions that individuals leave online [51].

There are several examples of the VGI: potential to forest-fire mapping [52], crisis-maps [53, 54], geotagged (Flickr) photograph analysis for tourism management [55–57], or mapping the sense of place [58]. Twitter is also an important source of data [59] and Takhteyev, Gruzd, and Wellman [60] studied the social ties between its users.

Additionally, Crampton et al. [61] evaluated the possible influences of big data on critical geography using exploratory methods to overcome some of the limitations related to the usage of VGI, and Viana et al. [62] accessed the value of OpenStreetMap (OSM) data for land use land cover (LULC) cartography. One may state that neogeography is bringing cartographic and GIS expertise to the common citizens [19]. Nevertheless, VGI properties are very different from the ones of traditional data sources. This can turn possible a more complex and dynamic interpretation than the one that census data allowed [63], but implies further research in the field of GISc [46].

Constantly, wide-ranging-data assembly allows to reverse engineer the events that stimulate the emergence of unexpected outcomes [27]. The complexity of geographic systems points for knowledge experimental analysis throughout the observation of processes [64]. The changing potential of big data is not about the size but instead about its spatiotemporal resolution, thematic coverage and omnipresence, and crossing analysis levels, from the single to the all [27, 42]. Now, we can study global patterns within geographic information networks [1].

Elwood, Goodchild, and Sui [46] define VGI as a "paradigmatic shift in how geographic information is created and shared" and reinforce the idea that further research is needed to develop new methods of spatial data analysis. Intrinsically, the advances we are seeing in the fields of VGI and/or big data open new research fields for GISc, especially regarding analytical capabilities. Simultaneously, GISc looks for a procedural background for dealing with the specific restrictions related to the usage of VGI and/or big data. Such challenges comprise the data quality and understanding in order to achieve a statistical valid sample, privacy issues and methods, and techniques for dealing with geographic data.

6. Conclusions

GISc is strongly connected Geography, as they equally analyze the same features of reality [65] using comparable outlooks. Thus, GISc is the "science behind the system" [5] mainly focused in computational and representation topics, while Geography aims to model and predict geographical phenomenon.

The dissatisfaction with traditional social physics (and geography) is comprehensible. They both looked for universal laws, which now, looking back, seems a little bit naıve to say the least. At the time, due to limited data and weak computation processing capacity, researcher in general and specifically geographers presumed homogeneity within physical and social systems. Doing that, we turn an exciting and data-rich environment, i.e., reality, into sterilized, amorphous, lifeless models. At their beginning, spatial analysis and GISc followed this approach despite having much more appealing representations, i.e., maps.

The new Geography, social physics, spatial analysis, and GISc are substantially different, as they are data and computation driven. The computer is the essential feature rather than an auxiliary one. When becoming more and more sophisticated, GISc assumes that generalization is possible although context is extremely important.

Lastly, instead of flattening geographic space into an insipid uniformity, GISc promotes heterogeneity as a key feature to understand how processes evolve and how to get better outcomes through a science-based policy. In addition, this continuous research can focus on the complexity of policy results. Social and social-physical systems are complex by nature and have particular dynamics with several feedback (positive or negative) loops. Some of these feedbacks are expected, like the mechanism that all living systems have for maintaining orderly conditions (i.e., homeostasis), and others are not, leading to the appearance of new system features (i.e., emergence). For that reason, it can be demanding (or even impossible) to evaluate the success of a policy intervention. Big data, and its related constant data assemblage, enable this to happen naturally, without constraints [3].

A policy intervention can result in numerous outcomes, positives or negatives, that will withstand for a while long [30]. The use of the actual time consuming deliberate experimental researches makes it problematic to explore all the amount of existing options [66]. Broader researchers' commitment to data-intensive analysis enables additional subtle, comprehensive, and profound approaches to complex

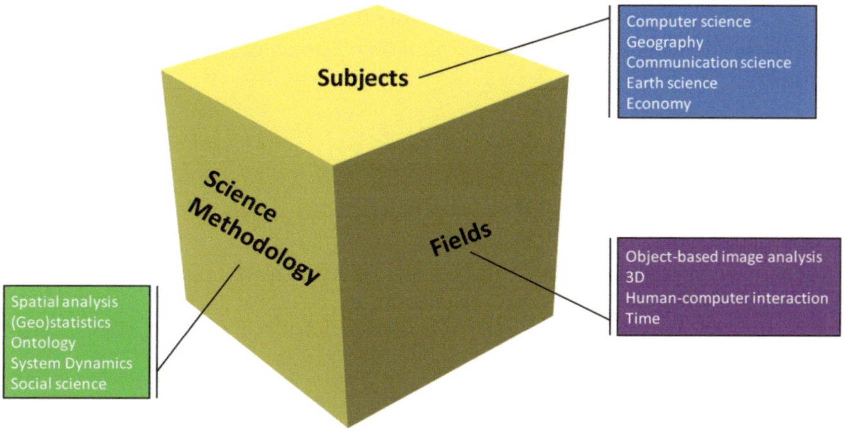

Figure 6.
Cubic representation of GISc research perspectives [73].

problems, refining the research and, at the same time, makes policies supported by science more understandable to the common people, e.g., climate change [67].

These progresses are still somewhat new [8], and GISc is just starting to methodically analyze whether these matters have (or not) the capability to leave a meaningful impact on society. Nonetheless, Big Data is probably the outmost important paradigm shift that, with more or less delay, will change GISc. The main question is the inherent and ever more spreading communicative status of geographic data [68, 69]. Web 2.0, VGI [70] and neogeography [15], and also the sensors that allow real time data capture prove that this a reality with no turning back [71].

Nevertheless, one should not be blind by data. Theory is critical to get enlightenments about what data reveal and for handling big data [42]. To avoid being trapped in a kind of data dependency, we need to understand the processes (including measurement methods and technologies) that generate it [72].

Blaschke and Merschdorf [7] recognized distinct trends in GISc. They systematized them into 10 items: (i) plenty of spatial data, (ii) thinking spatial, (iii) fuzzy analysis and turning into geographic nonspatial data, (iv) spatial computing, (v) ubiquitous computing, (vi) non-Cartesian measurements, (vii) spherical innovative spatial analysis, (viii) VGI, (ix) neogeography, and (x) geographic knowledge. From a citizen's and/or researcher's perspective, these 10 trends can be grouped into five main clusters: (i) big data and location analytics; (ii) the reborn of time geography with mobile users, mobile sensors, and trajectories; (iii) cognition, emotions, and other data unmeasurable in a straightforward manner; (iv) a more spatially aware society with geobrowers and/or virtual globes; and (v) the discovery of new geographic scales with, for instance indoor geospatial analysis.

The future will certainly continue to include various research fields (**Figure 6**), and yet, it is inspiring to enable harmony among processes and patterns. Looking to the possible interactions, one can isolate three groups, i.e., location analytics and mapping, spatiotemporal modeling, and social media and citizens. This last one clearly includes the user's perspectives, and it is heavily connected with the interdisciplinary that characterizes GISc.

Thinking and spatial reasoning constitute a form of thinking grounded in the concept of space, in the tools of representation and in the process of reasoning [74], and are stimulated through the manipulation of geotechnologies [75, 76]. Researches should act with extreme caution; as recently Facebook and Google shown to the world, there is not a crisp line, but rather a very fuzzy one, between naive and not-so naive social investigation [77]. The protection of citizens' privacy and the way it interlinks with the increasing need for data is a key point in the future of GISc [78].

Acknowledgements

We acknowledge the GEOMODLAB—Laboratory for Remote Sensing, Geographical Analysis and Modeling—of the Centre for Geographical Studies/Institute of Geography and Spatial Planning, University of Lisbon, for providing the required equipment and software.

Conflict of interest

The authors declare no conflict of interest.

Funding

This work was financed by national funds through FCT—Portuguese Foundation for Science and Technology, I.P., through Cláudia M. Viana Ph.D. grant (SFRH/BD/115497/2016) and under the framework of the project TRIAD—healTh RIsk and social vulnerability to Arboviral Diseases in mainland portugal (PTDC/GES-OUT/30210/2017).

Author details

Cláudia M. Viana, Patrícia Abrantes and Jorge Rocha
Institute of Geography and Spatial Planning, University of Lisbon, Lisbon, Portugal

*Address all correspondence to: jorge.rocha@campus.ul.pt

References

[1] Adams B, Gahegan M, Gupta P, et al. Geographic information observatories for supporting science. In: GIO 2014: Proceedings of the Workshop on Geographic Information Observatories 2014, Collocated with the 8th International Conference on Geographic Information Science (GIScience 2014). Austria: Vienna; 2014. pp. 32-39

[2] Janowicz K, Adams B, McKenzie G, Kauppinen T. Towards geographic information observatories. In: Proceedings of the Workshop on Geographic Information Observatories 2014, collocated with the 8th International Conference on Geographic Information Science (GIScience 2014) [Internet]. Vienna, Austria: CEUR; 2014. p. 1-5. Available from: http://ceur-ws.org/Vol-1273/GIO_intro.pdf

[3] Stanley K, Bell S, Kreuger LK, et al. Opportunistic natural experiments using digital telemetry: A transit disruption case study. International Journal of Geographical Information Science. 2016;**30**:1853-1872

[4] McKenzie G, Janowicz K. Information Observatories: What Are They Good For? In: Proceedings of Workshop on Geographic Information Observatories In conjunction with the Twelfth Conference on Spatial Information Theory (COSIT '15) [Internet]. 2015. p. 1-3. Available from: http://geog.ucsb.edu/~jano/io_gio2015.pdf

[5] Goodchild MF. Geographical information science. International Journal of Geographical Information Systems. 1992;**6**:31-45

[6] Reitsma F. Revisiting the 'Is GIScience a science?' Debate (or quite possibly scientific gerrymandering). International Journal of Geographical Information Science. 2013;**27**:211-221

[7] Blaschke T, Merschdorf H. Geographic information science as a multidisciplinary and multiparadigmatic field. Cartography and Geographic Information Science. 2014;**41**:196-213

[8] Goodchild MF. Geographic information systems and science: Today and tomorrow. Procedia Earth and Planetary Science. 2009;**1**:1037-1043

[9] Mark DM. Foundations of geographic information science. In: Duckham M, Worboys M, editors. Geographic Information Science: Defining the Field. New York: Taylor & Francis; 2003. pp. 3-18

[10] Gore A. The digital earth. Australian Survivor. 1998;**43**:89-91

[11] Goodchild MF. The future of digital earth. Annals of GIS. 2012;**18**:93-98

[12] Annoni A, Craglia M, Ehlers M, et al. A European perspective on digital earth. International Journal of Digital Earth. 2011;**4**:271-284

[13] Craglia M, de Bie K, Jackson D, et al. Digital earth 2020: Towards the vision for the next decade. International Journal of Digital Earth. 2012;**5**:4-21

[14] Zook M, Dodge M, Aoyama Y, et al. New Digital Geographies: Information, Communication, and Place BT—Geography and Technology. In: Brunn SD, Cutter SL, Harrington JW (eds). Dordrecht: Springer Netherlands; 2004. pp. 155-176

[15] Turner A. Introduction to Neogeography. O'Reilly Media. 2006. Available form: http://shop.oreilly.com/product/9780596529956.do

[16] Longley PA, Goodchild MF, Maguire DJ, et al. Geographic Information Science and Systems. 4th ed. Hoboken, NJ, USA: John Wiley & Sons; 2015

[17] Couclelis H. Climbing on a milestone for a better view: Goodchild's 'geographical information science' paper as vantage point and ground for reflection. International Journal of Geographical Information Science. 2012;**26**:2291-2300

[18] Kemp K, Kuhn W, Brox C. A Delphi survey to rate GIScience publication outlets. In: Multidisciplinary Research on Geographical Information in Europe and beyond Proceedings of the AGILE'2012 International Conference on Geographic Information Science. 2012. pp. 268-271

[19] Blaschke T, Donert K, Gossette F, et al. Virtual globes: Serving science and society. Information. 2012;**3**:372-390

[20] Chrisman N. Full circle: More than just social implications of GIS. Cartographica the International Journal for Geographic Information and Geovisualization. 2005;**40**:23-35

[21] Harvey FJ, Chrisman NR. The imbrication of geography and technology: The social construction of geographic information systems BT—geography and technology. In: Brunn SD, Cutter SL, Harrington JW (Eds). Dordrecht: Springer Netherlands; 2004. pp. 65-80

[22] Batty M. Building a science of cities. Cities. 2012;**29**:S9-S16

[23] Batty M. The New Science of Cities. Massachusetts: The MIT Press; 2013

[24] Bettencourt LMA. The origins of scaling in cities. Science. 2013;**340**:1438-1441

[25] Bettencourt L, West G. A unified theory of urban living. Nature. 2010;**467**:912

[26] Pentland A. Social Physics: How Good Ideas Spread—The Lessons from a New Science. London, UK: Penguin Press. Available from: https://books.google.pt/books?id=KvTKAgAAQBAJ; 2014

[27] Miller HJ, Goodchild MF. Data-driven geography. GeoJournal. 2015;**80**:449-461

[28] Conte R, Gilbert N, Bonelli G, et al. Manifesto of computational social science. The European Physical Journal Special Topics. 2012;**214**:325-346

[29] Barnes TJ, Wilson MW. Big data, social physics, and spatial analysis: The early years. Big Data & Society. 2014;**1**:2053951714535365

[30] Pollock K. Policy: Urban physics. Nature. 2016;**531**:S64

[31] McLafferty S. Spatial context. International Encyclopedia of Geography. Epub ahead of print. 2017. DOI: 10.1002/9781118786352.wbieg0705

[32] Wilson A. Entropy in urban and regional modelling: Retrospect and prospect. Geographical Analysis. 2010;**42**:364-394

[33] Wilson A. Entropy in Urban and Regional Modelling. Reprint (1. London, UK: Routledge. 2011. Available from: https://books.google.pt/books?id=0HTKq7GHZ4UC

[34] Fotheringham AS. Spatial interaction. International Encyclopedia of Geography. Epub ahead of print 6 March 2017. DOI: 10.1002/9781118786352.wbieg0749

[35] Walton D. Abductive Reasoning. Alabama: The University of Alabama Press; 2013

[36] DiBiase D, Demers M, Johnson A, et al. Geographic Information Science and Technology Body of Knowledge. 1st ed. Washington, D.C: Association of American Geographers; 2006

[37] Haklay M. Geographic information science: Tribe, badge and sub-discipline. Transactions of the Institute of British Geographers. 2012;**37**:477-481

[38] Frank AU. GIS theory—the fundamental principles in giscience: A mathematical approach. In: Harvey FJ, editor. Are There Fundamental Principles in Geographic Information Science. Tobler Lecture Event 2012. CreateSpace Independent Publishing Platform. 2012. pp. 12-41

[39] Chrisman NR. A Deflationary Approach to Fundamental Principles in Giscience. In: Harvey FJ (ed) Are There Fundamental Principles in Geographic Information Science. Tobler Lecture Event 2012. CreateSpace Independent Publishing Platform; 2012. pp. 42-64

[40] Remler DK, Van Ryzin GG. Research Methods in Practice: Strategies for Description and Causation. 2nd ed. London, UK: SAGE Publications, Inc; 2015

[41] Munro E. Evidence-based policy. In: Cartwright N, Montuschi E (eds) Philosophy of Social Science: A New Introduction. Oxford, England: Oxford University Press; 2014. pp. 48-67

[42] González-Bailón S. Social science in the era of big data. Policy & Internet. 2013;**5**:147-160

[43] Hendler J. Broad data: Exploring the emerging web of data. Big Data. 2013;**1**:18-20

[44] Tiropanis T, Hall W, Hendler J, et al. The web observatory: A middle layer for broad data. Big Data. 2014;**2**:129-133

[45] Goodchild MF. Citizens as sensors: The world of volunteered geography. GeoJournal. 2007;**69**:211-221

[46] Elwood S, Goodchild MF, Sui DZ. Researching volunteered geographic information: Spatial data, geographic research, and new social practice. Annals of the Association of American Geographers. 2012;**102**:571-590

[47] Hall W, Tiropanis T, Tinati R, et al. The web science observatory—The challenges of analytics over distributed linked data infrastructures. ERCIM News. 2014;**96**:29-30

[48] Tiropanis T, Hall W, Shadbolt N, et al. The web science observatory. IEEE Intelligent Systems. 2013;**28**:100-104

[49] Van Kleek M, Smith DA, Tinati R, et al. 7 billion home telescopes: Observing social machines through personal data stores. In: Proceedings of the 23rd International Conference on World Wide Web. New York, NY, USA: ACM; 2014. pp. 915-920

[50] Atzmanstorfer K, Blaschke T. The Geospatial Web: A Tool to Support the Empowerment of Citizens through E-Participation? In: Citizen e-Participation in Urban Governance: Crowdsourcing and Collaborative Creativity. Hershey, PA (USA): IGI Global; 2013. pp. 144-171

[51] Hall W, Tiropanis T. Web evolution and web science. Computer Networks. 2012;**56**:3859-3865

[52] Goodchild MF, Glennon JA. Crowdsourcing geographic information for disaster response: A research frontier. International Journal of Digital Earth. 2010;**3**:231-241

[53] Zook M, Graham M, Shelton T, et al. Volunteered geographic information and crowdsourcing disaster relief: A case study of the Haitian earthquake. World Medical & Health Policy. 2010;**2**:7-33

[54] Liu SB, Palen L. The new cartographers: Crisis map mashups and the emergence of neogeographic practice. Cartography and Geographic Information Science. 2010;**37**:69-90

[55] Encalada L, Boavida-Portugal I, Cardoso Ferreira C, et al. Identifying tourist places of interest based on digital imprints: Towards a sustainable smart city. Sustainability. 2017;**9**. Epub ahead of print. DOI: 10.3390/su9122317

[56] Encalada L, Ferreira CC, Rocha J, et al. Geographical patterns in the tourist city: GIS for spatiotemporal analysis. In: Chaudhuri S, Ray N, editors. GIS Applications in the Tourism and Hospitality Industry. Hershey, PA, USA: IGI Global; 2018. pp. 76-97

[57] Encalada L, Ferreira CC, Boavida-Portugal I, et al. Mining big data for tourist hot spots: Geographical patterns of online footprints. In: Koutsopoulos K, de Miguel González R, Donert K, editors. Geospatial Challenges in the 21st Century. Cham: Springer International Publishing; 2019. pp. 99-123

[58] Li L, Goodchild MF. Constructing places from spatial footprints. In: Proceedings of the 1st ACM SIGSPATIAL International Workshop on Crowdsourced and Volunteered Geographic Information. 2012. pp. 15-21

[59] Leetaru K, Wang S, Cao G, et al. Mapping the global twitter heartbeat: The geography of twitter. First Monday. 2013;**18**. Epub ahead of print. DOI: 10.5210/fm.v18i5.4366

[60] Takhteyev Y, Gruzd A, Wellman B. Geography of twitter networks. Social Networks. 2012;**34**:73-81

[61] Crampton JW, Graham M, Poorthuis A, et al. Beyond the Geotag: Situating 'big data' and leveraging the potential of the Geoweb. Cartography and Geographic Information Science. 2013;**40**:130-139

[62] Viana MC, Encalada L, Rocha J. The value of OpenStreetMap historical contributions as a source of sampling data for multi-temporal land use/cover maps. ISPRS International Journal of Geo-Information. 2019;**8**. Epub ahead of print. DOI: 10.3390/ijgi8030116

[63] Madaan A, Tiropanis T, Srinivasa S, et al. Observlets: Empowering analytical observations on web observatory. In: Proceedings of the 25th International Conference Companion on World Wide Web. Republic and Canton of Geneva, Switzerland: International World Wide Web Conferences Steering Committee. 2016. pp. 775-780

[64] O'Sullivan D, GLW P. Spatial Simulation: Exploring Pattern and Process. Chichester, England: John Wiley & Sons, Ltd; 2013. Epub ahead of print 2013. DOI: 10.1002/9781118527085

[65] Mark DM. Geographic information science: Defining the field. In: Duckham M, Goodchild MF, Worboys M, editors. Foundations of Geographic Information Science. London, UK: CRC Press; 2003. pp. 3-18

[66] Hausmann R. The Problem With Evidence-Based Policies. Project Syndicate: The Worlsd's Opinion Page. 2016. Available from: https://www.project-syndicate.org/commentary/evidence-based-policy-problems-by-ricardo-hausmann-2016-02?barrier=accesspaylog [Accessed: 23-04-2019]

[67] Townsend A. Cities of data: Examining the new urban science. Public Culture. 2015;**27**:201-212

[68] Crampton JW. Maps as social constructions: Power, communication and visualization. Progress in Human Geography. 2001;**25**:235-252

[69] Miller CC. A beast in the field: The Google maps Mashup as GIS/2. Cartographica the International Journal for Geographic Information and Geovisualization. 2006;**41**:187-199

[70] Flanagin AJ, Metzger MJ. The credibility of volunteered geographic information. GeoJournal. 2008;**72**:137-148

[71] Resch B, Mittlböck M, Girardin F. Live geography—embedded sensing for standarised urban environmental monitoring. International Journal on Advances in Systems and Measurements. 2009;**2**:156-167

[72] Lazer D, Kennedy R, King G, et al. The parable of Google flu: Traps in big data analysis. Science. 2014;**343**:1203-1205

[73] Blaschke T, Strobl J, Donert K. Geographic information science: Building a doctoral programme integrating interdisciplinary concepts and methods. Procedia—Social and Behavioral Sciences. 2011;**21**:139-146

[74] National Research Council [NRC]. Learning to Think Spatially. Washington, DC: The National Academies Press. Epub ahead of print; 2006. DOI: 10.17226/11019

[75] Jo I, Bednarz S, Metoyer S. Selecting and designing questions to facilitate spatial thinking. The Geography Teacher. 2010;**7**:49-55

[76] Bodzin AM, Fu Q, Peffer T. Investigating Curriculum Enactment with a GT-Supported Science Curriculum on Students' Geospatial Thinking and Reasoning. In: Bodzin K, Peffer T, editors. Association for Science Teacher Education (ASTE) Annual Meeting. Clearwater Beach, FL. 2012. pp. 1-23

[77] Vertesi J. The Real Reason You Should Be Worried About That Facebook Experiment. Time. 2014. Available from: http://time.com/2950699/facebook-experiment-social-science-funding/ [Accessed: 23-04-2019]

[78] Miller HJ, Tolle K. Big data for healthy cities: Using location-aware technologies, open data and 3D urban models to design healthier built environments. Built Environment. 2016;**42**:441-456

Section 2

Location Analytics and Modelling

Chapter 2

Evaluation of Nursing Facility Locations Using the Specialization Coefficient of the Population Aging Rate

Koya Tsukahara and Kayoko Yamamoto

Abstract

Recently, in Japan, the increase of aging population is especially rapid, the lack of nursing facilities has become a serious social issue, and the political measures against it are continuously enacted. Though the number of nursing facilities and its capacity are increasing, the utilization rate of such facilities remains at the same level, and the lack of facilities has not improved. Based on the background mentioned above, using geographic information systems (GIS) and public open data, the present study aims to quantitatively evaluate the current situation of nursing facility locations in urban areas within Japan. In the present study, the model of the p-median problem used to obtain the optimal location of facilities was modified, and a method to evaluate the current situation concerning the shortage or overage of nursing facilities by area was proposed. As evaluations are conducted using quantitative data, the evaluation results are also quantitative, making it an effective indicator for evaluating the locations of nursing facilities. Additionally, the specialization coefficient of the population aging rate and the distance between nursing facilities and areas were calculated based on public open data. Therefore, the evaluation method has a high temporal reproducibility as well as spatial reproducibility.

Keywords: nursing facility, aging population, p-median problem, facility location problem, geographic information systems (GIS), public open data

1. Introduction

While the aging population in many advanced countries around the world is increasing, the increase is especially rapid, and the lack of nursing facilities has become a serious social issue in Japan. According to the Ministry of Health, Labour and Welfare (MHLW) [1, 2], though the admission capacity of special nursing homes is approximately 498,000 as of 2014, the number of people requesting admission was 524,000. This grave situation highlights the severity of the lack of certified care workers as well as nursing facilities. In order to increase the number of certified care workers, the MHLW released a "Basic guideline concerning measures to increase workers in social welfare services" and is making efforts to secure personnel. Though the number of people registering as a certified care worker is increasing, the lack of workers remains an issue.

Meanwhile, though the number of nursing facilities and its capacity are increasing, the utilization rate of such facilities remains at the same level and the lack of facilities has not improved. Though subsidy is provided by the national and local governments for the construction of nursing facilities, the amount of subsidy cannot be greatly increased due to the need for more childcare and medical facilities as well, making it difficult to expect a great increase of nursing facilities in the future. Therefore, as a solution to the lack of facilities, the construction of new nursing facilities should be prioritized in areas with greater needs. In order to make this possible, first of all, it is necessary to accurately grasp the areas that lack nursing facilities.

Additionally, there are various facility types, and certain characteristics can be seen in selecting the location according to the type. For example, commercial establishments such as convenience stores are located in busy areas that are highly populated in order to get more customers and increase profit. On the other hand, educational and public facilities such as schools tend to be located in areas where many people can fairly access them. Similar to the latter case, instead of being concentrated in certain areas, nursing facilities must be located where everyone in need can access it in a fair manner. Based on the background mentioned above, using geographic information systems (GIS) and public open data, the present study aims to quantitatively evaluate the current situation of nursing facility locations in urban areas within Japan as the target.

2. Related work

The present study is related to A. studies concerning the sufficiency degree of nursing facilities, B. studies concerning care services provided from nursing facilities, and C. studies concerning facility location problems. The following will introduce the major preceding studies in the above three study areas and discuss the originality of the present study in comparison with the others.

In A. studies concerning the sufficiency degree of nursing facilities, as this topic attracts attention especially in Japan, there were a lot of preceding studies until now. As mentioned in Section 1, the reason for this is that the increase of aging population is especially rapid and the lack of nursing facilities has become a serious social issue in recent Japan. Yamada et al. (2008) [3] grasped the regional difference in demand for nursing facilities by means of interviews. Yamamoto et al. (2015) [4] considered elements such as the positioning of various equipment and nursing modality within the recovery rehabilitation ward as spatial structure and examined the reality of activities of residents in their leisure time. Acker et al. (2015) [5] assessed the nursing facility characteristics, quality ratings, and the views of facility administrators about the implications of an increasing number of foreign-born employees in Washington State in the USA. Takase et al. (2016) [6] grasped the reality of the operation of home care and organized the travel distance of caretakers as well as the service zone using the location and road information of users, caretakers, and offices in Japan. Fujita et al. (2017) [7] organized objections concerning the construction of nursing facilities and examined the changes in social awareness toward such conditions over time in Japan. Hunnicutt et al. (2018) [8] examined and quantified geographic variation in the initiation of commonly used opioids and prescribed dosage strength among older nursing home residents in the USA. Tahara et al. (2018) [9] clarified the attitude of social welfare facility staff regarding the acceptance of evacuees when natural disasters occur and elucidated factors that obstruct the acceptance of such evacuees.

For B. studies concerning care services provided from nursing facilities, there were a lot of preceding studies especially in the USA. Yun et al. (2010) [10] developed and validated an algorithm to identify the use of nursing facility services and differentiate short- from long-term care using Medicare claim data. Walsh et al. (2012) [11] analyzed potentially avoidable hospitalizations (PAHs) for dually eligible beneficiaries receiving long-term or post-acute care services to inform the development of health policies and better care programs and outcomes for this population. Cherubini et al. (2012) [12] examined resident and facility characteristics associated with hospitalization in a cohort of the older nursing home residents in Italy. Onder et al. (2012) [13] assessed the nursing home residents in Europe, focusing on the services and health for the elderly in Long TERm care (SHELTER) study. King et al. (2013) [14] examined how skilled nursing facility (SNF) nurses the transitional care of individuals admitted from hospitals, the barriers they experience, and the outcomes associated with variation in the quality of transitions. Neuman et al. (2014) [15] measured the association between SNF performance measures and hospital readmissions among Medicare beneficiaries receiving post-acute care at SNFs. Fry et al. (2018) [16] used robot cats to reduce the total number of falls in the facility quality improvement methods (strategy for improvement, design, setting, participants, interventions, measurements, and evaluation). Yamaguchi (2018) [17] grasped the relationships between received quality of care and information sharing among workers in nursing facilities for the elderly.

For C. studies concerning facility location problems, Segawa et al. (1996) [18] developed a system that can simulate factors related to childcare facility improvements, such as the extension of childcare hours and the location of new childcare facilities. Nagashima et al. (2014) [19] proposed an algorithm that derives the best location for electric vehicle (EV) power stations by means of the mean field approximation. Ozgen et al. (2014) [20] combined a two-phase possibilistic linear programming approach and a fuzzy analytical hierarchical process approach to optimize two objective functions (minimum cost and maximum qualitative factor benefit) in a four-stage (suppliers, plants, distribution centers, and customers) supply chain network in the presence of vagueness. Munemasa et al. (2015) [21] used the linear relaxation method to propose a method that derives the best solution for minimizing travel costs for the urban model made up of residential and business areas. Zhang et al. (2016) [22] investigated a facility location problem incorporating service competition and disruption risks, developing a new binary bilevel linear programming (BBLP) model. Ohdate et al. (2017) [23] considered relocation of facilities for the management of public facilities and categorized them based on building, function, and location to create an evaluation method for them. Nagai et al. (2017) [24] proposed an agent-based urban model in which the relationship between a central urban area and a suburban area was simply expressed. Usui et al. (2018) [25] theoretically investigated the relationship between the continuous walking distance distribution and the density of resting places.

On the other hand, in recent Japan, there are distinctive preceding studies that adopted an economic method into C. studies concerning facility location problems. Tanaka et al. (2015) [26] applied the quintile share ratio (QSR), which is an indicator showing the degree of bias in income, to the facility locational analysis for linear cities. Additionally, with QSR as a reference, Tanaka et al. (2016) [27] used the median share ratio (MSR), which is an equity measure, to develop a facility location evaluation model in a linear city with one or two facilities, as well as a uniformly distributed population. Furuta et al. (2017) [28] used a method that generalized the QSR and proposed a solution to optimize multiple facility locations in cases where the demand and candidate facility locations are discrete.

Regarding studies related to A, though the studies grasped the awareness of the local residents and awareness and behavior of users, as well as the actual condition of operation for nursing facilities, the location of the facilities was not considered. Regarding studies related to B, though the studies investigated the care services provided from nursing facilities, the location of the facilities was not considered. Regarding studies related to C, Nagashima et al. (2014) [19] and Munemasa et al. (2015) [21], respectively, considered EV power stations and business-and-residential distribution as their study subject and proposed a method to derive efficient locations. Though Segawa et al. (1996) [18] conducted simulations with the assumption that the facilities will be relocated, as there are currently many existing nursing facilities in cities of Japan, the above simulations cannot propose realistic solutions for such facilities. Additionally, though Tanaka et al. (2016) [27] and Furuta et al. (2017) [28] focused on the equity concerning the facility location evaluation method, it has only been applied to virtual cities and not to any actual cities. Therefore, with the results of the preceding studies mentioned above as a reference, the present study will demonstrate the originality by considering the lack of nursing facilities, which has become a serious social issue, and quantitatively evaluating current facility locations.

3. Evaluation framework and methods

3.1 Evaluation framework and process

For the evaluation method of nursing facility locations, PostgreSQL Ver. 9.6.1 and ArcGIS Pro Ver. 2.0 of Environmental Systems Research Institute (ESRI) were used. The evaluation framework and process are as mentioned below:

3.1.1 Creating the distribution maps of the aging population and nursing facilities as well as a road network map

The distribution maps of the aging population and nursing facilities as well as a road network map are created in digital map form using GIS. These three types of digital maps are superimposed, and the closest road node from a representative point from each nursing facility and area (divided by town and street) is set on the road network map.

3.1.2 Calculating the specialization coefficient of the population aging rate and adding it to the road network map

The specialization coefficient of the population aging rate is calculated using the aging population in addition to the total population data from each area, and the results are added to the road network map.

3.1.3 Calculating the shortest route using A algorithm*

By applying the data obtained in section 3.1.2 to A* algorithm, which is explained in the following section, the shortest route between each nursing facility and each area is calculated.

3.1.4 Evaluating nursing facility location using the specialization coefficient of the population aging rate

Using the distribution maps of the aging population obtained in section 3.1.1 and the shortest route calculated in section 3.1.3, the evaluation points are calculated for each area.

3.2 Evaluation method

3.2.1 Creating the distribution maps of the aging population and nursing facilities as well as road network maps

3.2.1.1 Distribution map of the aging population

The age group being evaluated in the present study is those over the age of 75. There are many cases where those over 65 can use nursing facilities. However, according to Hashimoto (2015) [29], the life expectancy and health span in Japan have become higher in recent year, with the latter being 71.19 for men and 74.21 for women. Therefore, assuming most users of nursing facilities are over 75, the age group was set for those over 75. As for the evaluation target area, in order to calculate evaluation points according to each area, GIS is used to display the distribution of the aging population on the digital map.

3.2.1.2 Distribution map of nursing facilities

GIS is used to display the distribution of nursing facilities on the digital map. While nursing facilities include facility types such as admission type, commuter type, and other related facilities, the present study will only consider admission type of nursing facilities.

3.2.1.3 Creating a road network map

As for the distance between nursing facilities and each area in the present study, the road network distance is used instead of the linear distance. This is because the linear distance may be extremely short compared to the distance when traveling along the roads to the nursing facilities, and this may cause the estimate of the travel distance of users to be shorter than it actually is.

First, GIS is used to display the road network map of the evaluation target area in digital map form. Next, the node closest to every nursing facility and the representative point in each area will be set up on the digital map. The representative point is the central point of the area, and the nodes are the intersections and endpoints of the roads. This is set up as the distance is calculated according to each node. In the present study, the node set as the representative point of the area is the start node, and the node set as the nursing facility is the end node.

3.2.2 Calculating the specialization coefficient of the population aging rate and adding it to the road network map

3.2.2.1 Introducing the specialization coefficient of the population aging rate

In the present study, the distance is weighted so that the road distance of areas with a large demand is longer, while the road distance of areas with a small demand

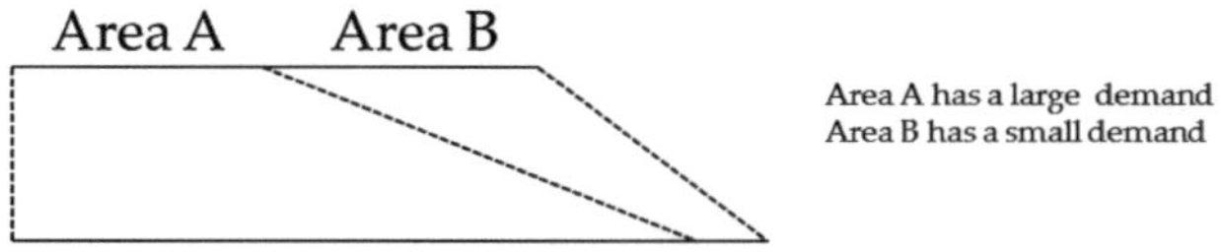

Figure 1.
Weighting pattern diagram.

is shorter, as shown in the weighting pattern diagram in **Figure 1**. If a user from a certain area selects a nursing facility that requires him or her to go through a transit area with a large demand, there is a high chance that there will be competition among users, as those from the transit area will most likely select the same nursing facility. Therefore, the weighting is conducted as mentioned above with the assumption that psychological distance of the users will be longer.

The specialization coefficient of the population aging rate is used as the weighting coefficient. The specialization coefficient of the population aging rate is the value indicating how high the population aging rate is in a certain area. Besides the specialization coefficient of the population aging rate, the population aging rate and the aging population can both be considered as a weighting coefficient. However, because the population aging rate is the ratio with the value always being 1 or less, the distance is always calculated to be shorter. Additionally, the difference in the aging population among the different areas is substantial enough to greatly affect the distance. On the other hand, the specialization coefficient of the population aging rate does not make the value too big or too small; hence, it is suitable as a weighting coefficient.

3.2.2.2 Calculating the specialization coefficient of the population aging rate

The specialization coefficient of the population aging rate in each area is calculated. In order to calculate the specialization coefficient, the population aging rate must be calculated first. The population aging rate and the specialization coefficient of the population aging rate are, respectively, calculated using Eqs. (1) and (2):

$$A_i = \frac{p_{75i}}{p_i} \tag{1}$$

where A_i is the population aging rate of area i (%), p_{75i} is the population of those over 75 in area i (people), and p_i is the total population of area i (people).

$$SC_i = \frac{A_i}{A} \tag{2}$$

where SC_i is the specialization coefficient of the population aging rate in area i, A_i is the population aging rate of area i (%), A is the population aging rate of all areas (%).

3.2.2.3 Adding the specialization coefficient of the population aging rate to the road network map

The specialization coefficient of the population aging rate calculated in Eq. (2) is added to the road network map using PostgreSQL. This is done by multiplying the distance between nodes by the specialization coefficient of the population aging

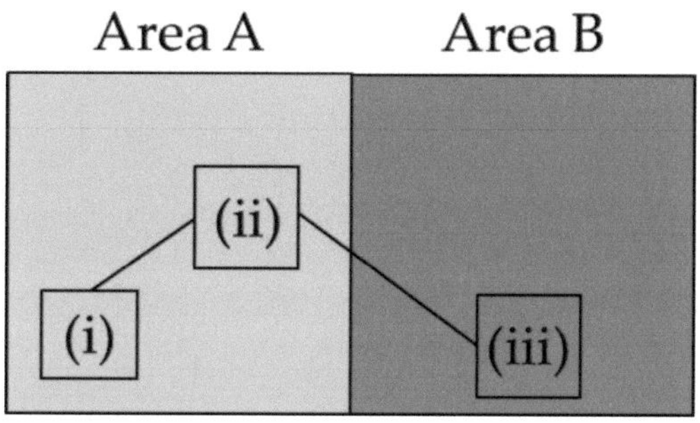

Figure 2.
Example of specialization coefficient of the population aging rate added to the road network map.

rate. As a rule for addition, the specialization coefficient of the population aging rate is added if the nodes are in the same area as in nodes (i) and (ii) of **Figure 2**. However, if the nodes are in different areas as in nodes (ii) and (iii), the average value of the specialization coefficient of the population aging rate is added.

3.2.3 Calculating the shortest route using A algorithm*

The shortest route between each nursing facility and each area will be calculated using A* algorithm. A representative search method for finding the shortest route is the Dijkstra method (Dijkstra, 1959) [30]. However, in the present study, the latitude and longitude are used for the coordinates of the representative points in each area as well as the nursing facilities, and the estimate value of the shortest routes between the two points are available in advance. Therefore, A* algorithm, which is the improved version of the Dijkstra method that can effectively calculate the shortest routes, is used.

In general, when considering the shortest route which starts at the start node, goes through node n, and ends at the end node, the route is expressed as shown in Eq. (3):

$$f^*(n) = g^*(n) + h^*(n) \tag{3}$$

where $f^*(n)$ is the shortest route distance (m), $g^*(n)$ is the shortest distance from the start node to n (m), and $h^*(n)$ is the shortest distance from n to the end node (m).

Though $f^*(n)$ can be easily obtained if the values of $g^*(n)$ and $h^*(n)$ are already known, in reality, the values of $g^*(n)$ and $h^*(n)$ are impossible to obtain beforehand. Therefore, the method of replacing $f^*(n)$ with the estimated $f(n)$ is called A* algorithm, and it is expressed as shown in Eq. (4):

$$f(n) = g(n) + h(n) \tag{4}$$

where $f(n)$ is the estimate value of the shortest route, $g(n)$ is the estimate value of the shortest route from the start node to n, and $h(n)$ is the estimate value of the shortest route from n to the end node.

An example of actually searching for the shortest route using A* algorithm on a computer is shown in **Figure 3**. In this example, the shortest route from coordinates (2,2) to (5,5) is obtained. The gray cells are set as impassable, while F, G, and H correspond to $f(n)$, $g(n)$, and $h(n)$ of Eq. (4). However, G is the actual travel distance from (2,2) to the current cell, and the distance for moving one cell over in

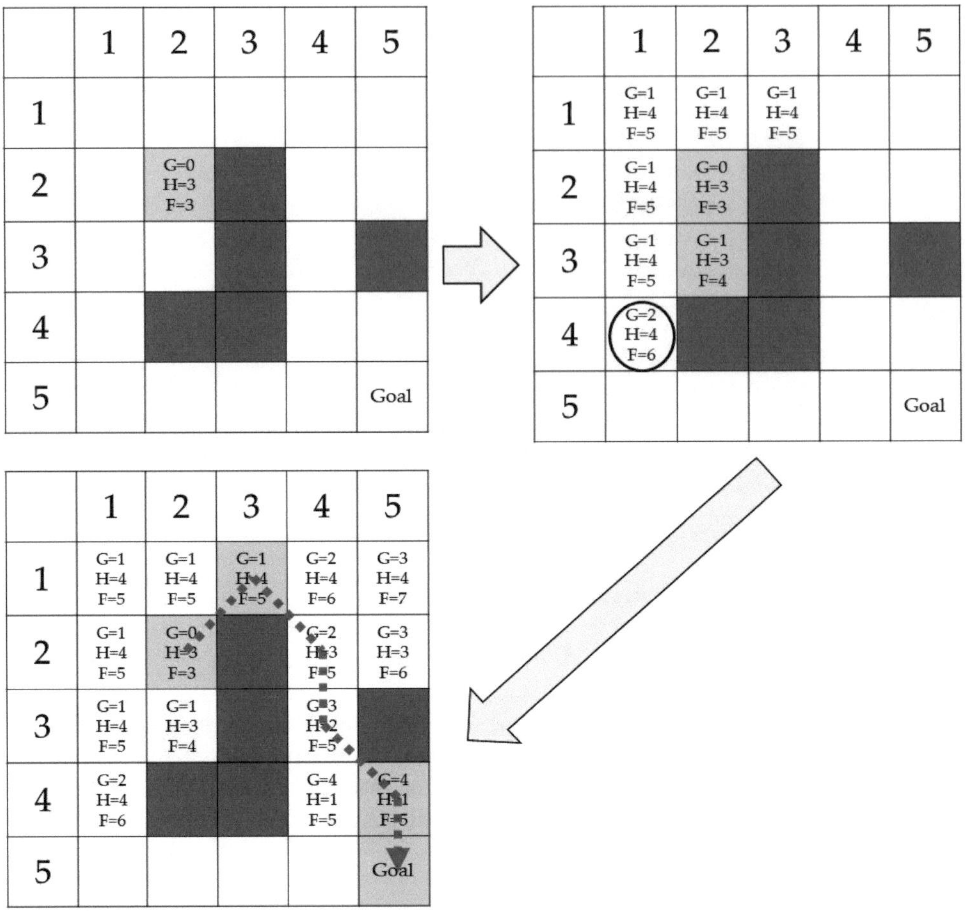

Figure 3.
Example of shortest route calculation using A algorithm.*

any direction is counted as 1. Additionally, for the estimate value of the shortest distance from the current cell to (5,5), H is set as the higher value of the difference between the x-coordinate and the y-coordinate of the two points.

The algorithm process is as follows:

1. Calculate the values of F, G, and H, surrounding the first cell.

2. Move to the cell with the value of F being the lowest.

3. Calculate the values of F, G, and H, surrounding the moved cell.

4. Among the cells newly calculated in the algorithm process 3, if the value of F is the same as or lower than that of the current cell, it is moved to that cell. If the value of F is higher than that of the current cell, the search on this route is discontinued. Then, among the value of cells already calculated, if the values of F, G, and H can be calculated and the value of F is the same or almost the same as that of the current cell, it is moved to this cell.

5. The algorithm processes 3 and 4 are repeated until the destination cell is reached.

The cells leading to the destination cell at this time is the shortest route, and the value of F obtained at the end is the distance of the shortest route. The shortest route in this example is indicated by the gray dotted line arrow, and the distance of the shortest route is 5. Though 1 was set as the distance when it was moved one cell over in the example of **Figure 3**, the distance can be changed according to each moved cell. In the present study, the distance of G is set as the product of the actual travel distance and the specialization coefficient of the population aging rate mentioned in Section 3.2.2.

3.2.4 Evaluation of nursing facility locations using the specialization coefficient of the population aging rate

The p-median problem is a facility location problem in which the total distance from the users to their closest facility is minimized, and this problem can be modeled as shown in Eq. (5). This model obtains the optimal location to reduce the load of users in all areas to the extent possible by changing the value of *Xij*:

$$\min_{X_{ij}} Z = \sum_{i}\sum_{j} w_i d_{ij} X_{ij} \mathrm{A_i} = \frac{\mathrm{P_{75i}}}{\mathrm{P_i}} \tag{5}$$

where $X_{ij} \in \{0, 1\}$. is the allocation for facility j in area i, w_i is the demand in area i, and d_{ij} is the distance from area i to facility j.

In the present study, the evaluation method is developed based on the p-median problem. Eq. (5) is a model that obtains the suitable location to reduce the travel load of users in all areas as much as possible and is different to the purpose of the present study. Therefore, in order to match the purpose of the present study, Eq. (5) is changed into Eq. (6):

$$\min_{d_{ij}} Z_i = w_i d_{ij} \mathrm{A_i} = \frac{\mathrm{P_{75i}}}{\mathrm{P_i}} \tag{6}$$

where Z_i is the evaluation point for area i, w_i is the demand in area i, and d_{ij} is the distance from area i to facility j (m).

Xij of Eq. (5) is removed in order for the facility location to stay the same, and by calculating the evaluation points according to each area, the facility location situation in each area can be quantitatively grasped. Additionally, as for the originality of the present study, distance *dij* is weighted by the specialization coefficient of the population aging rate as mentioned in Section 3.2.2. With *Dij* as the weighted distance, Eq. (6) can be expressed as Eq. (7):

$$\min_{D_{ij}} Z_i = w_i D_{ij} \tag{7}$$

where, Z_i is the evaluation point for area i, w_i is the demand in area i, and $D_{ij}$is the weighted distance of area i to facility j (m).

The evaluation points of each area is calculated after applying the aging population and the distance of shortest routes to nursing facilities of each area obtained in Sections 3.2.1 and 3.2.2, and the results are displayed on the digital map using GIS.

4. Selection of evaluation target area and data processing

4.1 Selection of evaluation target area

For the evaluation target area in the present study, Chofu City, Tokyo, was selected. Chofu City is located in the suburban area of Tokyo as shown in **Figure 4**.

Chofu City was selected as the evaluation target area, as the aging population (age 65 or over) has already exceeded the youth population (under age 15), and the aging population is expected to increase in the future. According to the survey on the aging population in Chofu City [31], 23,545 people, equivalent to approximately 10% of the total population, fit in the current age range of the present study subject which is 75 and over. In the present study, evaluation will be conducted in the unit of 105 areas within Chofu City.

4.2 Data processing

4.2.1 Data overview

The utilized data and the utilization method of the data in the present study are shown in **Table 1**.

4.2.2 Distribution map of aging population and nursing facilities

The distribution of the aging population and nursing facilities are, respectively, shown in **Figures 5** and **6**. As shown in **Figure 5**, the southeastern part has a high aging population, and there are old housing complexes in this area. In **Figure 6**, the admission type of nursing facilities is only shown as mentioned in Section 3.2.1, indicating that they are distributed throughout the entire Chofu City excluding the central area.

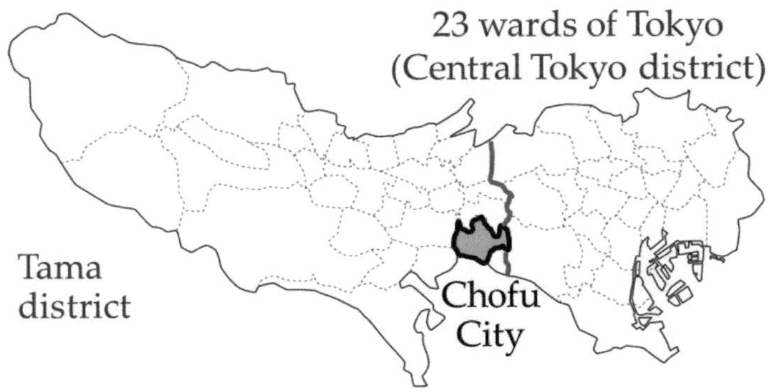

Figure 4.
Location of Chofu City in Tokyo.

Utilized data	Utilization method of data
Population by age (National Census 2010 by the Statistics Bureau)	Create the distribution map of the aging population Calculate the specialization coefficient of the population aging rate and add it to the road network map Calculate the shortest route using A* algorithm
Local care resources (Regional figures—Chofu City, Tokyo, by the Japan Medical Association)	Create the distribution map of nursing facilities
OpenStreetMap	Create the road network map Calculate the shortest route using A* algorithm

Table 1.
List of utilized data.

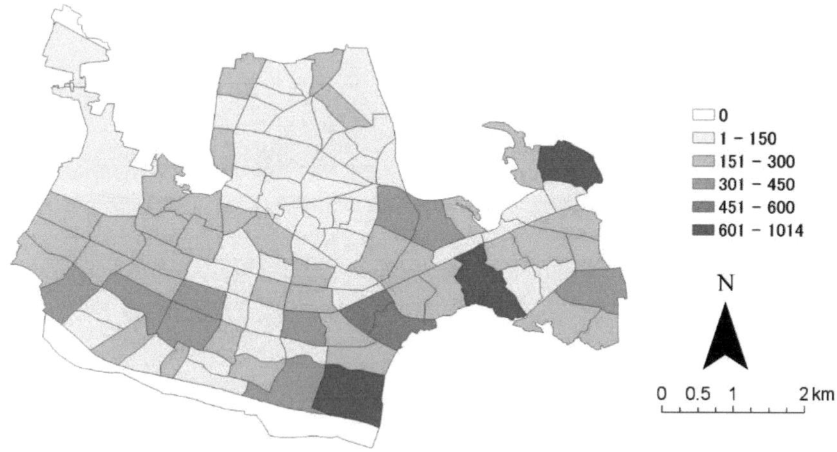

Figure 5.
Distribution of the aging population.

Figure 6.
Distribution of nursing facilities.

5. Evaluation

5.1 Evaluation for case with the specialization coefficient

Figure 7 is the evaluation results when using the specialization coefficient of the population aging rate. Areas with particularly high evaluation points are outlined by bold black lines. As evident in **Figure 7**, in addition to the areas with a high aging population mentioned in Section 4.2.2, evaluation points were also high in the eastern areas. On the other hand, evaluation points in the northern part of Chofu City were generally low.

5.2 Evaluation for case without the specialization coefficient

Figure 8 is the evaluation result without the use of the specialization coefficient of the population aging rate. Particularly high evaluation points are outlined by bold

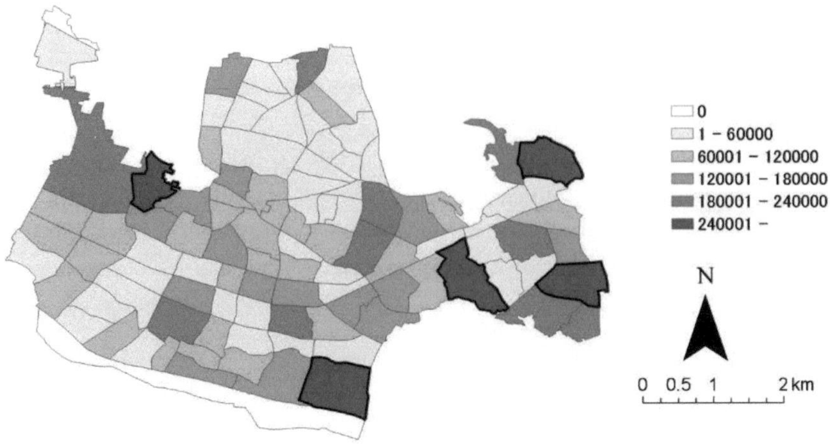

Figure 7.
Evaluation result when using the specialization coefficient.

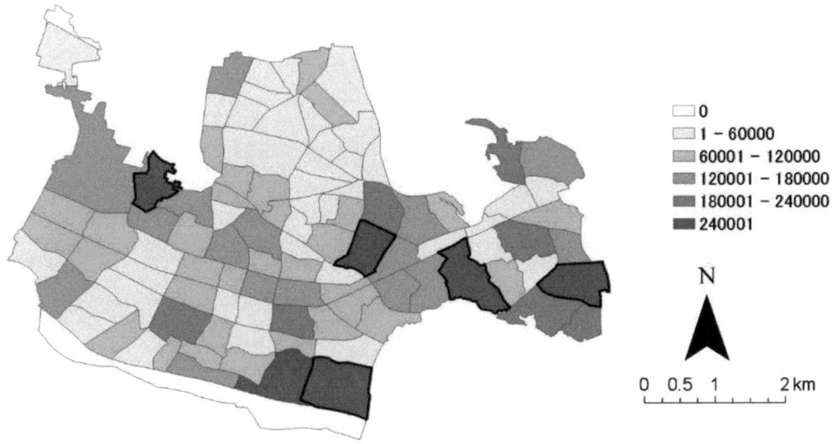

Figure 8.
Evaluation results when the specialization coefficient was not used.

black lines in the same manner as **Figure** 7, and among the areas outlined, four areas were the same as those outlined in **Figure** 7.

5.3 Comparing the evaluation results with and without the specialization coefficient

Figure 9 shows the difference between the evaluation results with and without the use of the specialization coefficient of the population aging rate. Areas with evaluation points higher in the former results are black, while areas with points higher in the latter are gray. Additionally, areas with particularly large or small differences are outlined with black bold lines. Regarding gray areas with high evaluation points when not using the specialization coefficient of the population aging rate, the evaluation points when using the specialization coefficient of the population aging rate were even higher, highlighting the lack of nursing facilities. With 32 areas marked gray and 25 marked black, it was made evident that the evaluation results when using the specialization coefficient of the population aging rate were lower overall.

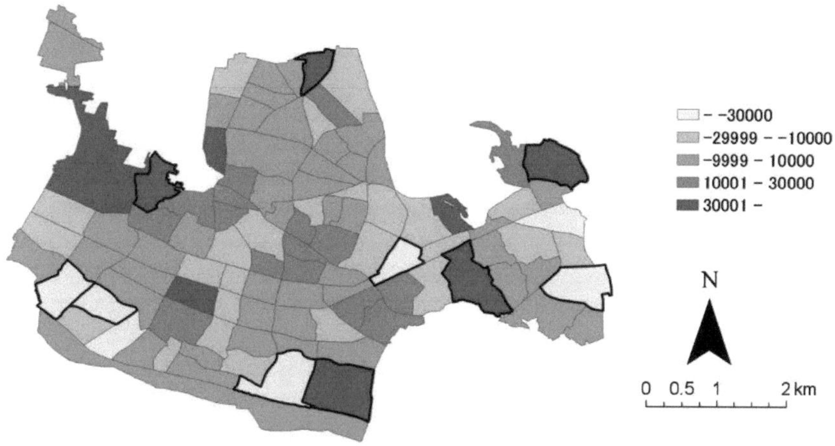

Figure 9.
The difference between evaluation results with and without the specialization coefficient.

5.4 Extraction of areas lacking nursing facilities

Based on the evaluation results of Section 5.1 (the evaluation results for cases with the specialization coefficient), the five areas with the highest evaluation points will be extracted as areas lacking facilities. The five areas outlined in light gray in **Figure 7** have a particularly high population of those 75 and over, and it can be said that there is currently a lack of nursing facilities. Additionally, as the population of those aged 65–74 are also high in these areas, further shortages of nursing facilities can be expected.

6. Conclusion

The conclusion of the present study can be summarized in the following three points:

1. In the present study, the model of the p-median problem used to obtain the optimal location of facilities was modified, and a method to evaluate the current situation concerning the shortage or overage of nursing facilities by area was proposed. By using the A* algorithm to calculate the distance between nursing facilities and areas, the shortest routes were efficiently calculated. Additionally, the specialization coefficient of the population aging rate used for the distance weighting can be used, because the users of nursing facilities are limited to the aging population, and the evaluation method which used this coefficient is a method specialized for nursing facilities.

2. Regarding the evaluation method in the present study, as evaluations are conducted using quantitative data such as the specialization coefficient of the population aging rate and the distance between nursing facilities and areas, the evaluation results are also quantitative, making it an effective indicator for evaluating the locations of nursing facilities. Additionally, as the current location situation of nursing facilities is evaluated by the area, the comparison of the shortage or overage of nursing facilities in each area as well as the extraction of areas lacking facilities can be easily conducted. Furthermore, as

the evaluation results are clearly displayed on the digital map using GIS, the shortage or overage of facilities can be visually understood.

3. In the present study, the specialization coefficient of the population aging rate and the distance between nursing facilities and areas were calculated based on public open data such as the National Census and OpenStreetMap. As evaluations are conducted based on public information, by obtaining population data and geospatial data similar to the present study, evaluations can be conducted using data in other areas as well as for the past and future. Therefore, the evaluation method in the present study has a high temporal reproducibility as well as spatial reproducibility. For example, by using the "future population estimate by region in Japan" of the National Institute of Population and Social Security Research as future data, the shortage or overage of nursing facilities in the future can be evaluated.

As research topics for the future, the improvement of the evaluation method for cases where there are multiple facilities within a certain range as well as the application of the evaluation method of the present study in the evaluation of other facility locations and other areas can be considered.

Conflict of interest

The authors declare that there is no conflict of interest that could be perceived as prejudicing the impartiality of the research reported.

Author details

Koya Tsukahara* and Kayoko Yamamoto
University of Electro-Communications, Tokyo, Japan

*Address all correspondence to: t1730063@edu.cc.uec.ac.jp

References

[1] Ministry of Health, Labour and Welfare: A Survey on Nursing Facilities and Offices.Tokyo: Ministry of Health, Labour and Welfare, 2016. 2p

[2] Ministry of Health, Labour and Welfare: Situation of the Applicants for Entrance to Nursing Facilities. Tokyo: Ministry of Health, Labour and Welfare, 2016. 3p

[3] Yamada A, Satoh E. Difference of management and nursing needs of small scale caring home for the elderly according to regionality. Journal of Architecture, Planning and Environmental Engineering. 2008; **73**(633):2355-2363

[4] Takase A, Yamada A, Nohara Y, Satoh E. Study on efficient placement and operation of home-visiting care in Provincial City. Journal of the City Planning Institute of Japan. 2016;**51**(3): 901-908

[5] Yamamoto T, Yanase R. A study on behavior of inpatients and spatial characteristic in convalescence rehabilitation wards. Journal of Architecture and Planning. 2015;**80** (714):1745-1751

[6] Acker K, Pletz AM, Katz A, Hagopian A. Foreign-born care givers in Washington state nursing homes: Characteristics, associations with quality of care, and views of administrators. Journal of Aging and Health. 2015;**27**(4):650-669

[7] Fujita Y, Saio N. Transition and problems of social consciousness about welfare facilities for the aged and children. Journal of Architecture, Planning and Environmental Engineering. 2017;**82**(733):697-703

[8] Hunnicutt JN, Baek J, Alcusky M, Hume AL, Liu SH, Ulbricht CM, et al. Geographic variation in the initiation of commonly used opioids and dosage strength in United States Nursing Homes. Medical Care. 2018;**56**(10):847-857. DOI: 10.1097/MLR.0000000000000972

[9] Tahara M, Kitagawa K. Takayama: A study on the function of social welfare facilities as evacuation centers. Japanese Journal of Social Welfare. 2018;**53**(1): 16-28

[10] Yun H, KilgoreJeffrey ML, Curtis JR, DelzellLisa E, Gary LC, Saag KG, et al. Identifying types of nursing facility stays using medicare claims data: An algorithm and validation. Health Services & Outcomes Research Methodology. 2010;**10**(1–2):100-110

[11] Walsh EG, Wiener JM, Haber S, Bragg A, Freiman M, Ouslander JG. Potentially avoidable hospitalizations of dually eligible medicare and medicaid beneficiaries from nursing facility and home- and community-based services waiver programs. Journal of the American Geriatrics Society. 2012;**60** (5):821-829

[12] Cherubini A, Eusebi P, Dell'Aquila G, Landi F, Gasperini B, Bacuccoli R, et al. Predictors of hospitalization in Italian Nursing Home residents: The U. L.I.S.S.E. project. Journal of the American Medical Directors Association. 2012;**13**(1):84.e5-84.e10

[13] Onder G, Carpenter I, Finne-Soveri H, Gindin J, Frijters D, Henrard JC, et al. Assessment of Nursing Home Residents in Europe: The Services and Health for Elderly in Long TERm care (SHELTER) study. BMC Health Services Research. 2012;**12**(5). DOI: 10.1186/1472-6963-12-5

[14] King BJ, Gilmore-Bykovskyi AL, Roiland RA, Polnaszek BE, Bowers BJ, Kind AJH. The consequences of poor

communication during transitions from hospital to skilled nursing facility: A qualitative study. Journal of the American Geriatrics Society. 2013;**61**(7): 1095-1102

[15] Neuman MD, Wirtalla C, Rachel M, Werner RM. Association between skilled nursing facility quality indicators and hospital readmissions. Journal of the American Medical Association. 2014;**312** (15):1542-1551

[16] Fry L, Walker M, Ledesma S, Gluch K, Deason K. Use of robotic cats to reduce falls in skilled nursing facility. Journal of Post-Acute and Long-Term Care Medicine. 2018;**19**(3):B22-B23. DOI: 10.1016/j.jamda.2017.12.070

[17] Yamaguchi I. Causal relationships between perceived quality of care and information sharing among workers in care facilities for the elderly. Japanese Journal of Communication Studies. 2018;**46**(2):131-149

[18] Segawa S, Sadahiro Y. A decision-making support system for child care facilities using GIS. Theory and applications of GIS. 1996;**4**(1):11-18

[19] Nagashima T, Shinzato T, Kaku I. Facility location problem of quick electric recharging system for electric vehicles. Journal of Japan Industrial Management Association. 2014;**64**(4): 557-560

[20] Ozgen D, Gulsun B. Combining possibilistic linear programming and fuzzy AHP for solving the multi-objective capacitated multi-facility location problem. Information Sciences. 2014;**268**:185-201

[21] Munemasa Y, Honda Y, Imai K. Constrained optimum arrangement structure which minimizes travel cost of residence and jobs. Journal of Architecture and Planning. 2015;**80** (712):1373-1380

[22] Zhang Y, Snyderc LV, Ralphs TK, Xue Z. The competitive facility location problem under disruption risks. Transportation Research Part E: Logistics and Transportation Review. 2016;**93**:453-473

[23] Ohdate S, Tsunoda M, Yi S. A study on phased optimization of the gross area of public facilities via facility relocation: Verification of the evaluation process and case studies in Machida-City. Journal of Architecture and Planning. 2017;**82**(732):363-370

[24] Nagai H, Kurahashi S. Bustle changes the city: Facility for stopping off and modeling urban dynamics. Transactions of the Japanese Society for Artificial Intelligence. 2017;**32**(1):D-G26_1-D-G2610

[25] Usui H, Hino K. Continuous walking distance and criteria for density of resting places: Application to Tokyo Central Station and Otemachi Station. Journal of Architecture and Planning. 2018;**83**(748):1049-1056

[26] Tanaka K, Furuta T. Quintile share ratio in a linear city with one and two facilities. Journal of the City Planning Institute of Japan. 2015;**50**(3):628-635

[27] Tanaka K, Furuta T. Evaluating inequality of facility location in a linear city using median share ratio. Journal of the City Planning Institute of Japan. 2016;**51**(3):894-900

[28] Furuta T, Tanaka K. Minimizing quantiles share ratio in multiple facility location problem with total distance constraint. Transactions of the Operations Research Society of Japan. 2017;**60**:36-49

[29] Hashimoto S. A Study on the Index Value Calculation of Healthy Life Expectancy An Examination of the Healthy Life Expectancy Regarding Promotion of National Health

Promotion Exercise (Health Japan 21) in the 21st Century; 2015. pp. 26-38

[30] Dijkstra EW. A note on two problems in connexion with graphs. Numerische Mathematik. 1959;**1**(1): 269-271

[31] Chofu City: A Survey on the Aging Population in Chofu City [Internet]. 2018. Available from: http://www.city.chofu.tokyo.jp/www/contents/1515569448760/index.html [Accessed: August 23, 2018]

Chapter 3

Machine Learning-Based Method for Urban Lifeline System Resilience Assessment in GIS*

Wenjie Huang and Mengzhi Ling

Abstract

System resilience, the capability of a system to sustain and recover from deliberate attacks, accidents, or naturally occurring threats or incidents, is a key property to measure the degree of robustness and coupling effect of complex system. The systems of waste disposal, urban water supply, and electricity transmission are typical systems with complex and high coupling features. In this chapter, a methodology for measuring the system resilience of such systems is proposed. It is a process of integrated decision-making which contains two aspects: (1) a five-dimensional indicator framework of system resilience which includes attributes in infrastructural, economic, and social sectors and (2) a hybrid K-means algorithm, which combines entropy theory, bootstrapping, and analytic network process. Through utilizing real data, the methodology can assist to identify and classify the level of system resilience for different geographical regions which are sustained by lifeline systems. The calculation of algorithm, visualization of processed data, and classification of resilience level can be finally realized in geographic information system. Through utilizing by regional governments and local communities, the final result can serve to provide guideline for resource allocation and the prevention of huge economic loss in disasters.

Keywords: system resilience, geographic information system (GIS), analytical network process, K-means

1. Introduction

Urban lifeline systems typically consist of critical infrastructure systems such as water, gas, electricity, communication, and transportation [1, 2]. In the modern era, these systems serve to maintain the function and ensure the quality of life for cities and their citizens. With the rapid urbanization and economic development, the demand of cities is increasing, and different regions of urban areas are interconnecting. This draws attentions to government and communities about the performance of lifeline system under the stresses of emergent status in disasters. Systematical methodologies and techniques, thus, are sought by the government and communities for identifying, analyzing, and prioritizing the resilient capacity

---

*Substantial part of this chapter was earlier published as a journal article: Huang, Wenjie, and Mengzhi Ling. 2018. "System resilienc assessment method of urban lifeline system for GIS." Computers, Environment, and Urban Systems 71:67–80.

of cities. The report of the United Nations Office for Disaster Risk Reduction indicated that "$366 billion in direct damages and 29,782 fatalities worldwide" are caused by natural disasters in 2011. This signifies that enhancing the resilient ability of urban lifeline systems can be a way of ensuring massive savings through risk reduction and expeditious recovery. However, there is currently no precise platform to analyze and visualize system resilience to support urban planning; furthermore, there lacks systematical methodologies for assessing multidimensional complex lifeline system resilience. This chapter proposed an integrated decision-making process for evaluating and classifying the system resilience levels of the urban lifeline systems of urban areas. It is a methodology based on weighted multi-criteria indicator data. This methodology can serve as guidelines for governments and communities to effectively reallocate resources for preventing economic loss and to conduct infrastructure reconstruction in those low resilience level areas. Particularly, the contribution of this chapter has four main parts:

1. For evaluating the system resilience systematically, a multi-attribute indicator framework of system resilience is integrated. This framework has been proposed in literature, which includes five-dimensional attributes.

2. For classifying the resilience level of urban regions, a hybrid K-means algorithm based on weight-adjusted data is proposed.

3. For effectuating the algorithm and visualizing the clustering results of resilient level, the GIS platform is introduced.

This chapter is structured by four sections. First, the literature review is embedded in the Introduction. Second, the methodology is elaborated and interpreted. Third, the application of the proposed methodology is illustrated through a sample case of salt tide hazard in China. Fourth, an overview of the contribution and applicability of this study and recommendation for further study are given.

1.1 Literature review

1.1.1 Resilience of complex systems

According to the Presidential Policy Directive (PPD)-21 on Critical Infrastructure Security and Resilience, the terminology "resilience" represents "the ability of a system to prepare for and adapt itself to changing conditions, and rapidly withstand and recover from disruptions." For urban systems, the domain of disruptions includes deliberate attacks, accidents, or naturally occurring threats or incidents [3]. In order to assist our cities to resist in and recover from disasters, many studies are focusing on establishing system resilience assessment. For instance, Francis and Bekera [25] proposed an assessment framework for system resilience which consists of five parts: "system identification, vulnerability analysis, resilience object setting, stakeholder engagement and resilience capacities." They also developed a triangle system resilience structure which includes "three pillars" of resilient capacity: absorptive, adaptive, and recovery ([4], p. 92). According to their definitions, absorptive capacity is the degree at which a system can absorb the impacts of perturbation and minimize consequences with little efforts ([4], p. 94). Adaptive capacity is the ability of a system to adjust to undesirable situations by undergoing minor changes ([4], p. 94). Recovery capacity is characterized by system reliability and the rapidity of the system returning to normal or improved operations ([4], p. 94).

City is an organism that consists of the physical environment and human beings. The interactions between the environment and human manifest the complexity of urban systems, that is, the mutual coupling among the environmental system and economic system, social systems, and other systems. Classical studies of system resilience, however, fail to take this complexity and attendant uncertainty of urban system into account [4]. In addition to evaluating the quality of lifeline infrastructure and the diversities in socioeconomic and cultural characteristics, the interactions between physical and nonphysical systems can also significantly contribute to the resilience of urban systems [5, 6].

Through studying 100 cities worldwide, an international construction firm, Arup, proposed the City Resilience Index (CRI) [7]. CRI is an urban resilience assessment framework structured by triple-level which consists of 4 dimensions (e.g., health, well-being, economy, and society), 12 goals (e.g., minimized human vulnerability and sustainable economy), and 52 indicators (e.g., protection of livelihoods following a shock and well-managed public finances) [7].

The assessment of complex system resilience enables decomposing resilience into its individual attributes and then organizing the attributes into an organizational tree of indicators from different aspects and levels. Accordingly, resilience could be evaluated through a combination of these indicators [8]. This provides ideas for collecting data from various sub-systems of urban organism, but a complex methodology concerning the interdependencies of such variables is required [3, 9, 10].

Pregenzer [11] believes that, in future nuclear system resilience research, new methodologies for soliciting expert opinions and analyzing historical data will be required to assess the relative strengths and potential unintended consequences of nonproliferation strategies. In a recent research by [12], for studying the long-lived socio-ecological systems in a causal loop diagram, multi-methods are applied. The methods include observation, expert opinion, and archive research. However, existing resilience assessment framework such as CRI is generally dependent on the method of expert scoring. Possible subjective bias caused by the method may reduce the reliability of the assessment results and have negative influence on further implement.

1.1.2 Resilience evaluation methods

The ability of a system to adequately and efficiently respond to external risks and internal instability is crucial in the context of urbanization and socioeconomic development. It is particularly critical to integrated information and communication systems and factors [9, 13]. For urban infrastructure systems, research has been done on developing generalized models to analyze the interdependencies among systems. One of the methods is called multilayer infrastructure network framework where economic flows and information flows transmit between different system layers. Accordingly, the system interdependency could be derived by computable general equilibrium (CGE) theory [10]. Mathematical modeling dealing with economic impact and recovery time is most commonly applied for resilience of supply chain system and related infrastructure. In make-to-stock systems, integral of the time absolute error is an appropriate control engineering measure of resilience for system inventory level and shipment rates [14]. Static assessment of system resilience helps to reflect intrinsic properties of certain lifeline systems. The work [15] proposes a qualitative approach to measure the resilience of residential buildings in various exogenous hazards scenarios, by using different parameters in the loss function and recovery function. Simulation techniques such as Monte Carlo analysis are used to conduct numerical studies on resilience evaluation and estimation [15].

1.1.3 Analytical network process

In the areas of risk assessment and decision analysis, analytical network process (ANP) or analytical hierarchy process (AHP) is widely used to assess the key factors of risks and analyze the impacts and preferences of decision alternatives [16]. The work [17] realizes the dynamic analysis of interaction among five urban elements: nature, society, economy, technology, and management and evaluation of regional flood hazard resilience through adopting ANP. However, subjective judgments of personnel and the scorings of decision-makers are common limitations of ANP and AHP. The ordered weighted average approach is a countermeasure to the constraints. It assists to eliminate these bias and subjective through making a series of local aggregations at each level of AHP [18, 19]. More advanced theories such as stochastic AHP transform the preference of decision-makers from stochastic pairwise comparisons to certain probability distributions with minimal information loss and gain optimal strategies by solving the problem of nonlinear programming model [20]. Other indicator weighting methods include robustly and stochastically weighted multi-objective optimization model [21], interval judgment matrix [22], and stochastic dominance [23].

Based on literature review, there exist several shortcomings in current system resilience evaluation methodologies. First, the mathematical definition of "system resilience" is absent which causes the constraint to measure system resilience quantitatively and elaborately. It means that in the study of system resilience assessment, instead of relying on a fixed mathematical model, new techniques purely based on real data are required. Second, there lacks a method to effectively integrate the human knowledge of decision-makers and experts into the process of system resilience evaluation. Accordingly, the methodology proposed in this chapter aims to improve the evaluation of system resilience in three aspects: (1) formulating an indicator pool that integrated factors in existing resilience assessment, (2) involving human knowledge to the process through inviting evaluators to select proper indicators from the pool for specific scenario of lifeline system and disaster (e.g., the water supply system facing salt tide and the electricity generation system experiencing hurricane) and to score, and (3) introducing techniques of machine learning to reduce the bias effects of human interpretations.

2. Methodologies

This section elaborates the proposed methodology for system resilience assessment of lifeline system. The methodology aims to identify and classify the resilience levels of urban regions supported by the lifeline system. For instance, it can be used to evaluate the resilience levels of residential areas, commercial areas, and industrial parks supported by the urban water supply system. This methodology has five parts:

1. An integrated system resilience indicator framework: [24] collected 36 system resilience assessment frameworks and proposed five dimensions of resilience assessment—material and environmental resources (M&ER), society and well-being (S&WB), economy (E), built engagement and infrastructure (BE&I), and governance and institutions (G&I) [24]. The proposed study adapted result in [24] as a resilience evaluation framework and called it MSEBG framework.

2. Adopting three pillars of system resilience capacity: the absorptive, adaptive, and recovery capacities from [25]. The evaluators, that is, the experts, then

chose the appropriate indicators from the above five dimensions under each pillar according to different hazards (e.g., the experts consider that "diverse population composition" is significant for the recovery of water supply systems during flood; therefore, they will classify this indicator under the category of "recovery capacity").

3. Using ANP model to assign the weights of the indicators. Explicitly, the weights represent the relative importance of each indicator from decision-maker's perspective.

4. Developing a hybrid K-means algorithm for adjusting weight and clustering: the algorithm proposed in this chapter combines standard K-means algorithm with the entropy theory and bootstrapping method. The algorithm uses real data of the indicator as input and then identifies and classifies the resilience level of different regions supported by the lifeline system.

5. Building GIS for calculation and visualization of resilience clustering results.

In **Figure 1**, the connection between various methods applied in this study is shown. In this methodology, the indicator framework will be built according to the selected appropriate indicators for measuring the resilience of specific system, and the raw data of indicators will be assigned according to the "three pillars" of capacities. Decision-makers use ANP model to generate their weights. The raw data and the weights are the inputs of the hybrid K-means algorithm; the algorithm contributes to cluster all the communities into groups with different resilience levels. These result data are finally transferred into a GIS platform; and the visual results can be shown.

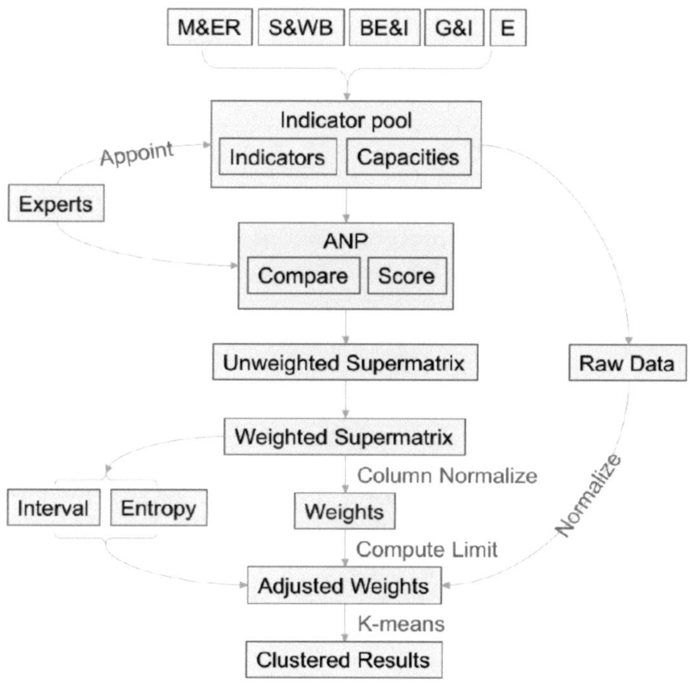

Figure 1.
Decision-making flow.

2.1 System resilience evaluation indicator framework

In general existing urban resilience assessment frameworks consist of different dimensions of urban system and various indicators for evaluation. In this chapter, the indicator framework directly applies the indicator matrix proposed by [24], who collected attributes of urban resilience from 36 assessment frameworks and categorized them into 5 dimensions with 122 indicators. Subsequently, indicators from these dimensions are classified into the "three pillars" of system resilience by evaluators. This will guide decision-makers to identify and extract key indicators for resilience assessment of the complex lifeline systems.

After construction of framework, the experts can select the detailed indicators from the indicator pool proposed by [24], classify the indicators into "three pillars" of capacities, and then derive the indicators' interdependence. A two-layer ANP model is constructed based on these indicators, in which the top layer is the "three pillars" of capacities and the bottom layer is constructed by the chosen indicators.

The experts choose an integer from −10 to 10 to represent the relative importance of any two indicators according to their perspective (for comparing "protection of wetlands and watersheds" and "availability and accessibility of resources," negative value means that "protection of wetlands and watersheds" is more important than the latter; and positive value means that "availability and accessibility of resources" are more important than the previous). The absolute value reveals the relative importance of an indicator inside a pair, i.e., selecting −10 and 10 means that one indicator strongly dominates and is dominated by the other, respectively, while 0 means they are almost equally important.

The ANP algorithm strictly refers to decision-making process introduced by [26]. The ANP will then generate weighted and unweighted super-matrices from the expert scoring. The row and column number of unweighted super-matrices equals to the number of indicators. The weighted super-matrices store the weights of indicators under the same capacity category and the weights between capacities. Explicitly, the weighted super-matrices are computed by the column normalization of unweighted super-matrices. The weight for all the indicators equals to the limited matrix iteratively computed and convergent from weighted super-matrices.

2.2 Hybrid K-means algorithm

The hybrid K-means algorithm is developed in this section: the overview of the algorithm is presented first. The interpretations at each step of this algorithm are presented later. The contents for the algorithm are shown as follows:

- N: the total number of all regions
- R: the total number of indicators
- $D \in \mathbb{R}^{N \times R}$: the column-normalized indicator data with N regions and R indicators. Column normalized means that each original element in the column is divided by the maximum value in that column. The calculation enforces each element in the matrix D taking value inside $[0, 1]$ and thus normalizes all the data into the same scale. Normalization is a standard procedure before performing K-means algorithm, when the indicators are of incomparable physical units.
- $Ws \in \mathbb{R}^{R \times 1}$: the weight of each indicator by the ANP model

- $SR \times R$: the initial super-matrix generated by the ANP model with R indicators
- a_{ij}, $i, j \in \{1, \ldots, R\}$: each element in the matrix S for a given super-matrix S
- $H \in \mathbb{R}^{1\times R}$ and L_i, $U_i \in \mathbb{R}^{R\times 1}$, $i \in \{1, \ldots, R\}$: entropy values
- $[L_i, U_i]$, $i \in \{1, \ldots, R\}$: closed intervals

The contents of hybrid K-means algorithm are summarized in Algorithm 1.

In Algorithm 1, Step 1a computes a confidence interval for each indicator based on bootstrapping method. Step 1b computes the entropy value from the super-matrix S. In Step 2, the optimal weight $\hat{Wo}$ equals to the optimal solution of the optimization problems (1)–(4), where the weight confidence interval constraint is incorporated as (3). In (4), to ensure that each weight is greater than zero, we set $\epsilon = 0.001$. In Step 3, Ws represents the subjective weight, and Wo represents the objective weight. Step 4 outputs the weight-adjusted indicator data. Finally, Step 5 implements the classical K-means algorithm. The structure of K-means algorithm follows what have been proposed in [27]. Set the number of clustering groups as k. After implementing Algorithm 1 and getting the clustering results, the term m_j, $j = 1, 2, \ldots, k$ is used to denote the centroid of each clustering group. The norm of each cluster centroid is computed as $C_j = \left\| m_j^{(T)} \right\|_2, j = 1, 2, \ldots, k$ and is then ranked from low to high, that is, $C_{(1)} \leq C_{(2)} \leq \ldots C_{(k)}$. Identifying the system resilience level of points in clustering group is based on the norm of the clustering centroid. The resilience level of the group with the centroid $D_{(j)}$ equals j.

2.2.1 Entropy and bootstrapping

In statistics, bootstrapping methods refer to the tests or metrics that rely on random sampling with replacement. Bootstrapping method allows assigning measures of accuracy (bias, variance, confidence intervals, prediction error, etc.) to sample estimates [28, 29]. In this chapter, Algorithm 2 shows the bootstrapping method that outputs the confidence interval for the weight of each indicator depending on super-matrix S in ANP and thus provides the feasible interval for the weight of each indicator for further adjustment and optimization.

The feasible interval of each indicator i is computed by resampling n samples from the initial sample space $\{a_{i1}, a_{i2}, \ldots, a_{iR}\}$, and then repeat such procedure for B times. Then compute the average value of the samples in the resampled space for each $b = 1, 2, \ldots, B$. Step. 2 and 3 in Algorithm 2 rank the average value of resample batches and then compute the estimated value for the 1–α confidence interval of the average value. The feasible interval of each indicator equals the estimated confidence interval, i.e., $[L_i, U_i] = \left[\hat{\theta}_i^{*,(k1)}, \hat{\theta}_i^{*,(k2)}\right]$. Further, the optimized models (1)–(4) with entropy objective are interpreted. Entropy theory has been applied to wide range of system resilience assessments from engineering and economics to anthropology and social ones [30–33]. Entropy indicates the degree of disorder, uncertainty, or lack of information about the configuration of system modules [34]. The lower the entropy value, the higher is the information utility it has. Here, the entropy value H computed by Algorithm 3 follows the definition in [35–37] and evaluates the information utility and reliability for the weights of indicators generated by ANP.

Algorithm 1. Hybrid K-means.

Input: Original indicators data $D \in \mathbb{R}^{N \times R}$; Weight matrix by ANP model $Ws \in \mathbb{R}^{R \times 1}$, and Super matrix $S \in \mathbb{R}^{R \times R}$;

Step 1: for $i, j = 1, \ldots, r$ **do.**

Step 1a: Compute the bounds $[L_i, U_i]$, $\in \{1, \ldots, R\}$ by row data $\{a_{i1}, a_{i2}, \ldots, a_{iR}\}$ $i = 1, \ldots, R$ from S using bootstrapping methods, summarized in *Algorithm 2*;

Step 1b: Compute the Entropy value H, based on column data $a_{1j}, a_{2j}, \ldots, a_{Rj}$ from S, summarized in *Algorithm 3*;

Step 2: Define $Wo \in \mathbb{R}^{R \times 1}$ with each element as w_i, $i \in \{1, \ldots, R\}$, given a positive threshold $\epsilon > 0$, and solve the linear program

$$\min_{Wo} \quad H \cdot Wo \tag{1}$$

$$\text{s.t.} \quad \sum_{j=1}^{R} w_i = 1, \tag{2}$$

$$L_i \leq w_i \leq U_i, \qquad i = 1, \ldots, R, \tag{3}$$

$$w_i \geq \epsilon, \qquad i = 1, \ldots, R, \tag{4}$$

which returns the optimal $\hat{W}o = \{\hat{w}_i\}$, $i = 1, 2, \ldots, R$.

Step 3: Denote $Ws_{(r \times 1)} = \{w'_i\}$, $i = 1, 2, \ldots, R$. Compute the adjusted weights by

$$W = \left\{ \frac{\hat{w}_i \times w'_i}{\sum_{i=1}^{R} \left(\hat{w}_i \times w'_i \right)} \right\}, i = 1, \ldots, R, \tag{5}$$

Step4 :*Construct weight matrix*$W' \in \mathbb{R}^{N \times R}$by*N copies of array*$W^{\top}$. *Compute the weight*

adjusted indicator data by Hadamard product of weight matrix W' and data matrix D, i.e., $D' = W' \circ D$

Step 5: Implement K-means algorithm on D'

Return clustering result on N regions;

Algorithm 2. Bootstrapping

Input: Define an initial sample space $\{a_{i1}, a_{i2}, \ldots, a_{iR}\}$ for each $i \in \{1, 2, \ldots, R\}$ from the Weighed supermatrix S; Confidence level α;

Step 1: Randomly select B batches of samples with size n from initial sample space, for each, $i \in \{1, 2, \ldots R\}$,

$$A_i^{*,b} = \left(a_{i1}^{*,b}, a_{i2}^{*,b}, \ldots, a_{in}^{*,b} \right), b = 1, 2, \ldots, B,$$

for $b = 1, 2, \ldots, B$ **do**, compute $\hat{\theta}_i^{*,b} = \frac{1}{n} \sum_{j=1}^{n} a_{ij}^{*,b}$, for $i \in \{1, 2, \ldots R\}$, and $b = 1, 2, \ldots, B$.

Step 2: Sort the value of $\hat{\theta}_i^{*,1}, \hat{\theta}_i^{*,2}, \ldots, \hat{\theta}_i^{*,B}$ from low to high, and obtain $\hat{\theta}_i^{*,(1)} \leq \hat{\theta}_i^{*,(2)} \leq \ldots \leq \hat{\theta}_i^{*,(B)}$, for each $i \in \{1, 2, \ldots R\}$.

Step 3: Set $k_1 = \alpha \times B/2$, and $k_2 = B - \alpha \times B/2$, let $\hat{\theta}_i^{*,(k_1)}, \hat{\theta}_i^{*,(k_2)}$ be the estimation of $\hat{\theta}_i^{*,\alpha/2}, \hat{\theta}_i^{*,1-\alpha/2}$, that is, the estimation of $1 - \alpha$ confidence interval of statistics θ.

Algorithm 3. Computation of Entropy.

Input: Super matrix $S = \{a_{ij}\} i = 1, \ldots, R, j = 1, \ldots, R$.

Step 1: Normalize the matrix element by $a'_{ij} = \frac{a_{ij}}{\sum_{j=1}^{R} a_{ij}}, \quad i, j \in \{1, \ldots, R\}$,

Step 2: Normalize by column to obtain $f_{ij} = \frac{a'_{ij}}{\sum_{i=1}^{R} a'_{ij}}$.

Step 3: for $j = 1,\ldots R$ **do**, compute entropy by definition:

$H_j = -\frac{1}{\ln(r)} \sum_{i=1}^{r} f_{ij} \ln\left(f_{ij}\right)$.

Return Entropy value H.

For S, each column vector $\left(a_{1j}, a_{2j}, \ldots, a_{Rj}\right)^{\mathrm{T}}, j = 1, 2, \ldots, R$ denotes the weighting strategy under the jth indicator's criterion; therefore, the weighted aggregation of each column's output entropy represents the total uncertainty metric of the indicator system. Summarized by optimization models (1)–(4), the algorithm seeks to find the optimal weight strategy $\hat{W}o$, to minimize the weighted aggregation entropy value $H \cdot Wo$, with respect to the feasible interval computed from bootstrapping confidence interval. Namely, the algorithm intends to find an optimal weight strategy to maximize the overall information utility. Such strategy would help to eliminate the side effects of subjective judgments from the experts. Superimposing the subjective and objective weights together realizes a comprehensive weight assignment strategy.

3. Case study

Case study of system resilience evaluation for a water supply system in the risks of salt tide is presented in this section. The case study validates the proposed methodology.

3.1 Background information

This section evaluates the Chenhang reservoir water supply system resilience under the salt tide hazards in the estuary of Yangtze River. The formal definition of salt tide is the emergency situation that chloride concentrations in water body exceed the national standard level (250 mg per liter of water). Salt tide destroys the quality of water, results in soil salinization in coastal areas and cities, and has negative impacts on production and human daily life. Recently, salt tide has already become one of the most internationally concerned disasters of coastal cities. The invasion of saltwater during salt tide will limit the access of high-quality municipal and industrial water and will cause water shortage and scheduling problems in some megacities of China located in estuarine and coastal areas, such as Shanghai. Thus, water shortage problem has become one of the main obstacles that obstructs the construction of eco-city and sustainable development. By experience, the intensity of salt tide intrusion changes with the period of tides, showing its periodic properties. In general, September to April next year is the period of time influenced by salt tide of water intake in Yangtze River. Each intrusion of salt tide lasts for 6–7 days. Since there exist multiple factors that affect the duration and extent of hazards (e.g., Yangtze River hydrology, chase traffic, weather, and wind), it is usually difficult to make detailed and accurate prediction for each intrusion. In recent years, the hazards have become more severe in the following aspects: long intrusion duration, high frequency, short interval time between intrusions, and independence of Yangtze River runoffs. Given the fact that the "Three Gorges Project" increased the ability to implement different water strategies at the upstream of Yangtze River, the extreme hydrological hazards occurred more frequently, which make the research on how salt tide influences the water supply system more practical and crucial [38–40].

According to chemistry knowledge, physical and chemical contaminates such as chloride concentration, ammonia, oxygen consumption, and total iron play a role as indicators to show the degree of salt tide level. These indicators reach peak values during December to February [41]. Accordingly, the whole period can be categorized into three different periods:

- Period I: September to December, when the chloride concentration in water accumulates up to 250 mg/L
- Period II: December to February (next year), when the chloride concentration abruptly reaches approximately the peak value (250–450 mg/L)
- Period III: From February to April, when salt tide recedes and chloride concentration returns to normal level

The goal of this case study is to study the system capacities in terms of preparation for salt tide hazards, facing salt tide hazards, and recovering from salt tide hazards. Period I is the preparation period, when the salt tides gradually affect the water supply system. In Period III, the recovery phase would occur, after the severe disturbances of salt tide. Intuitively, for Period I, decision-makers should focus more on increasing the adaptive capacity of system when prevention and emergency are crucial to system against increasing salinity. In Period II, absorptive capacity should be emphasized, since the system needs to absorb perturbations and minimizes the consequence when contaminants immediately reach the peak. In Period III, recovery capacity is more important indicator when water supply system needs redesigning and rebuilding. Thus, for each period, the decision-making process can be implemented dynamically, and the weight evaluations of indicators and clustering results can be derived step by step.

In the studied area, Chenhang reservoir supplies water to districts (e.g., Baoshan, Jiading, Putuo, Zhabei, and Hongkou) and towns (e.g., Gaodong, Gaoqiao, and Gaoxing inside Pudong New District) in Shanghai. The whole study region is zoned into 29 communities on the basis of administrative divisions shown in **Figure 2**. A unique FID number is assigned to each community.

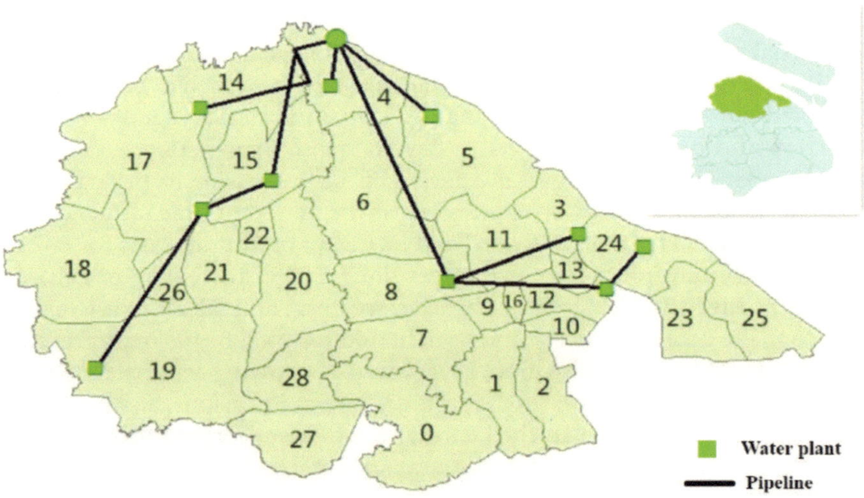

Figure 2.
Basic GIS Layer for Chenhang reservoir water supply system.

3.2 Resilience evaluation results

The resilience evaluation results are presented and analyzed in this subsection.

3.2.1 ANP model

In Section 2, the MSEBG indicator framework for system resilience and ANP model were exhaustively introduced. The indicators selected from the experts for this case study are shown in **Table 1**.

In this case study, 50 experts were involved. Those experts are chosen based on the selection method proposed in [42]. Half of the experts are professional technicians from lifeline system industry or experts in the field of system resilience, and another half of the experts are part-time college students majoring in Environmental Engineering and Urban Design. With solid technical and practical backgrounds,

	Absorptive capacity
M & E R	Protection of wetlands and watersheds
	Availability and accessibility of resources (water)
S & WB	Degree of connectedness across community groups
E	Investment in green jobs and green economy
BE & I	Robustness and fortification
	Spatial distribution of critical infrastructure
	Regular monitoring, maintenance, and upgrade of critical infrastructure
G & I	Decentralized responsibilities and resources
	Understanding risk patterns and trends
	Adaptive capacity
M & E R	Material and resource management
S & WB	Place attachment and sense of community and pride
E	Income and poverty
	Stability of prices and incomes and property value
BE & I	Redundancy of critical infrastructure, facilities, and stocks
	Consolidation of critical utilities and collaboration between utility providers
	Diverse and reliable information and communication technology (ICT) networks
G & I	Strong leadership
	Drills and exercises
	Recovery capacity
M & E R	Ecosystem monitoring and protection
	Protection of wetlands and watersheds
S & WB	Population composition
	Past experience with disaster recovery
E	Insurance and social welfare
G & I	Efficient management of resources

Table 1.
Indicators for water supply system.

they can understand and evaluate system resilience and provide scores for the corresponding indicators.

The ANP model for this case study is presented in **Figure 3**. The connection arc shows the independent relationship between two indicators.

3.2.2 Results and analysis for clustering

The data of the selected indicators (denoted by D in Algorithm 1) comes from *2011 Statistical Yearbook of Shanghai* issued by Shanghai Municipal Statistics Bureau (SMSB) and 2011 statistical yearbooks issued by Statistic Bureau of Shanghai. The corresponding semantic-type or non-quantifiable indicator data are transformed into real number value by expert scoring method. As summarized in Section 2, the input data D is preprocessed with column normalization to eliminate the bias effects.

Figure 4 illustrates the system resilience levels of all communities in the region under three periods. Particularly, all the districts are clustered into four resilience levels: Level 1 (particularly low), Level 2 (relatively low), Level 3 (relatively high),

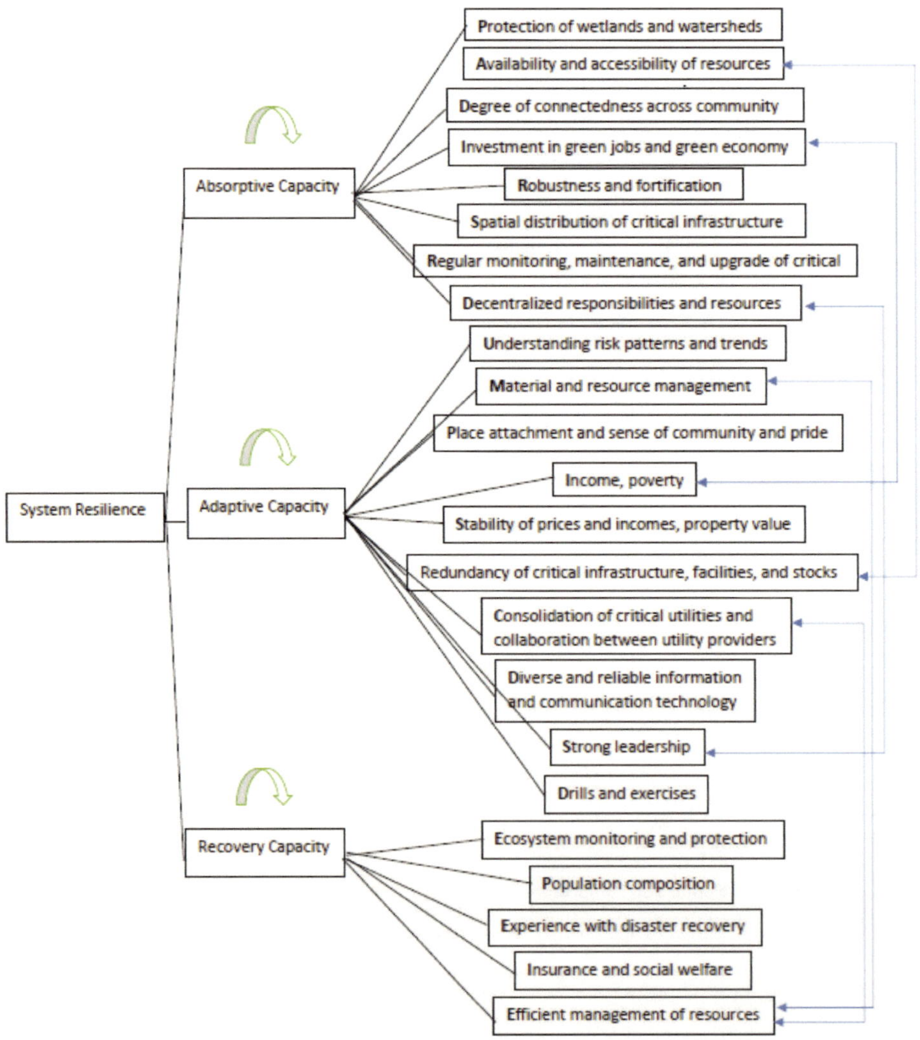

Figure 3.
ANP model for water supply system.

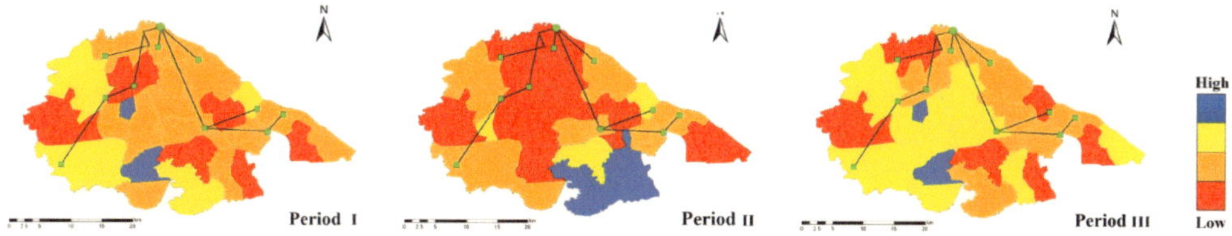

Figure 4.
System resilience level under Periods I, II, and III.

and Level 4 (high). The four levels are represented by colors red, orange, yellow, and blue. Such four level clustering concepts are inherited from warning signals of meteorological disasters on [43]. The visualized clustering result from GIS system enables decision-makers to prioritize crucial prevention and protections to the areas with low system resilience level and act dynamically when water supply systems face different periods of salt tide. **Table 2** shows that the resilience levels of most of the communities are clustered into Level 1 and Level 2. **Figure 4** shows a

Period I			Period II		Period III	
j	D_j	Region unit FID	D_j	Region unit FID	D_j	Region unit FID
1	0.8990	2, 7, 11, 15, 18, 23, 26	0.9353	4, 6, 9, 11, 14, 15, 16, 18, 20, 21, 22, 23, 25	0.9083	2, 7, 11, 14, 18, 23, 26
2	1.0175	1, 2, 4, 5, 6, 8, 9, 10, 12, 13, 14, 16, 20, 21, 24, 25, 27	1.2046	5, 8, 10, 12, 13, 17, 19, 24, 27	1.0811	1, 4, 5, 6, 7, 8, 9, 10, 12, 13, 15, 16, 20, 21, 24, 25, 27
3	1.2638	0, 3, 17, 19	1.3689	3, 7, 26	1.2039	0, 3, 17, 19
4	1.7721	22, 28	1.8976	0, 1, 2	1.1610	22, 28

Table 2.
Overall system resilience level analysis results under Periods I, II, and III.

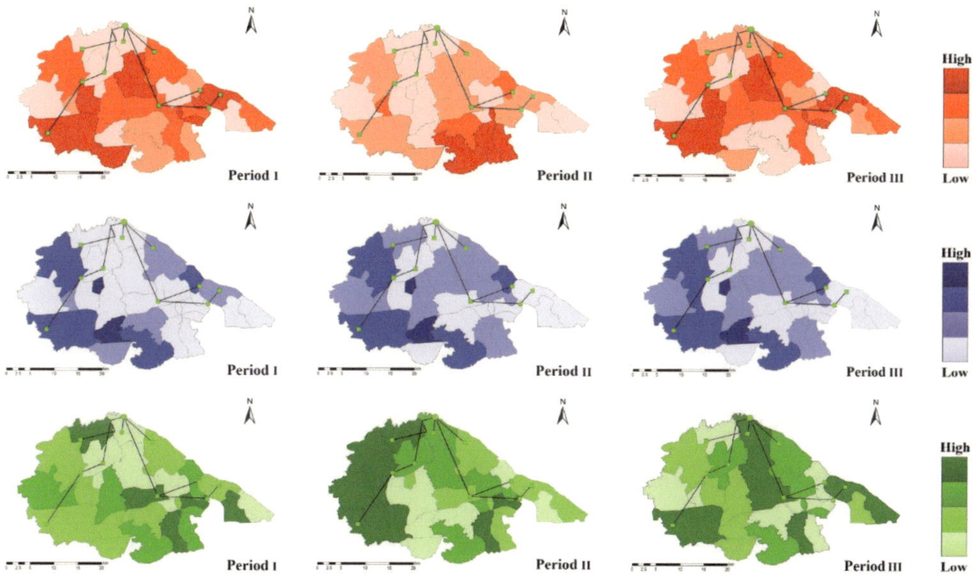

Figure 5.
Absorptive (red), adaptive (blue), and recovery (green) capacities under Periods I, II, and III.

phenomenon that the communities closer to the reservoir and contained inside the network of the water supply systems have lower resilience levels. These areas show closer exposure to the hazard and have less time to provide rapid responses, and also these communities take charge of the protection of network in disaster; reversely, the communities located far away from the reservoir, as well as the downtown area, have better economic status and higher service level. Absorptive, adaptive, and recovery capacities are analyzed, respectively, for Periods I, II, and III, with the results expressed in **Figure 5**. The complete clustering results are presented in **Table 3** with associated FID. **Figure 5** shows that for different periods, the resilience level of each community varies between three different capacities.

Period I				Period II		Period III	
Capacity		***D***	**Region unit FID**	D_j	**Region unit FID**	D_j	**Region unit FID**
Absorptive	1	0.6163	4, 11, 14, 15, 18, 22, 23, 28	0.9267	4, 9, 10, 11, 14, 15, 18, 20, 21, 22, 23, 25, 28	0.2692	4, 11, 14, 15, 18, 22, 23, 28, 26
	2	0.7354	0, 2, 3, 7, 26	1.1995	5, 6, 8, 10, 12, 13, 17,19,24,27	0.4120	0, 2, 3, 7
	3	0.7988	1, 5, 8, 9, 12, 17, 20, 25	1.3633	3, 7, 26	0.4597	1,5,8,9,12,17,20,25
	4	0.9418	2, 3, 6, 10, 13, 16, 19, 21, 24, 27	1.8941	0, 1, 2, 16	0.5284	2, 3, 6, 10, 13, 16, 19, 21, 24, 27
Adaptive	1	0.4996	1, 2, 4, 6, 8, 9, 10, 11, 12, 13, 14, 15, 16, 18, 20, 21, 23, 25	0.0669	4, 7, 8, 9, 10, 12, 15, 16, 21, 23, 24, 25, 27	0.1216	4, 7, 8, 9, 10, 12, 15, 16, 21, 23, 24, 25, 27
	2	0.7357	5, 7, 24, 26, 27	0.0894	1, 2, 5, 6, 11, 13, 14, 18, 20, 26	0.1693	1, 2, 5, 6, 11, 13, 14, 18, 20
	3	0.8538	0, 3, 17, 19	0.1589	0, 3, 17, 19	0.3134	0, 3, 17, 19, 26
	4	1.6285	22, 28	0.3322	22, 28	0.6941	22, 28
Recovery	1	0.3043	4, 6, 9, 11, 15, 22, 25	0.0529	3, 7, 20, 21, 23, 27, 28	0.8393	2, 3, 7, 9, 12, 13, 14, 18, 26
	2	0.3119	5, 7, 13, 16, 17, 19, 20, 21, 24, 27, 28	0.0573	2, 5, 8, 10, 16, 22, 24, 25	0.9826	17, 20, 21, 23, 27, 28
	3	0.3295	0, 2, 3, 10, 13, 18, 20, 26	0.0607	0, 4, 6, 9, 11, 13, 15	0.9858	0, 5, 10, 11, 15, 16, 22
	4	0.3460	1, 8, 12, 14, 23	0.0667	1, 12, 14, 17, 18, 19, 26	1.1447	1, 4, 6, 8, 19, 24, 25

Table 3.
Three capacity analyses under Periods I, II, and III.

4. Conclusion

The objective of this study was to propose a decision-making process and a methodology of system resilience assessment for urban lifeline systems. In this work, formation of the concept "system" should not only be limited to system infrastructures but also be expanded to the combination of other related complex systems such as demographic system, economic system, and environmental system in the study region. The advantages of using ANP as weight methods for the indicators come from two aspects: first, ANP structures the decision-making process by considering both the hierarchy relationship and the interdependence between bottom level indicators; and that is what "networks" in "ANP" comes from; and second, ANP generates weights through sequential pairwise comparisons of experts after selecting any two of the indicators, which is a very straightforward approach for real-life application. There exist four advantages for using hybrid K-means algorithm: (1) the number of clustering groups can be set before running the whole process; (2) the performance of K-means algorithm in high-dimensional clustering problems is relatively superior than other clustering algorithms such as fuzzy C-means, mountain, subtractive, hierarchical, and density-based clusterings in terms of quality, accuracy, and computation time [44–46]; (3) it provides the information of central points of each clustering class, which enables decision-maker to compute their distance from the origin and conduct further spatial analysis and (4) it is a machine learning technique, rather than by a formal mathematical metric of "resilience". Thus this model-free method can be implemented without an explicit mathematical definition of "resilience".

Author details

Wenjie Huang[1*] and Mengzhi Ling[2]

1 Department of Industrial Systems Engineering and Management, National University of Singapore, Singapore

2 Department of Architecture, School of Design and Environment, National University of Singapore, Singapore

*Address all correspondence to: wenjie_huang@u.nus.edu

References

[1] Moteff J, Copeland C, Fischer J. Critical infrastructures: What makes an infrastructure critical? Library of Congress Washington DC Congressional Research Service; 2003

[2] Zhao X, Cai H, Chen Z, Gong H, Feng Q. Assessing urban lifeline systems immediately after seismic disaster based on emergency resilience. Structure and Infrastructure Engineering. 2016;**12**(12): 1634-1649

[3] Ayyub BM. Systems resilience for multihazard environments: Definition, metrics, and valuation for decision making. Risk Analysis. 2014;**34**(2): 340-355

[4] Fiksel J. Designing resilient, sustainable systems. Environmental Science & Technology. 2003;**37**(23): 5330-5339

[5] Horne AL, Haynes RW. Developing measures of socioeconomic resiliency in the interior Columbia River basin; 1999

[6] Official Website of the World Bank-Data-Indicators; http://data.worldbank.org/indicator

[7] RF TRF, Arup. City resilience index: Understanding and measuring city resilience; 2017. http://www.arup.com/cityresilienceindex

[8] Renschler C, Frazier A, Arendt L, Cimellaro GP, Reinhorn A, Bruneau M. Developing the 'PEOPLES' resilience framework for defining and measuring disaster resilience at the community scale. In: Proceedings of the 9th US National and 10th Canadian Conference on Earthquake Engineering (9USN/10CCEE); Toronto. 2010. pp. 25-29

[9] Helbing D. Globally networked risks and how to respond. Nature. 2013; **497**(7447):51-59

[10] Zhang P, Peeta S. A generalized modeling framework to analyze interdependencies among infrastructure systems. Transportation Research Part B: Methodological. 2011;**45**(3):553-579

[11] Pregenzer A. Systems Resilience: A New Analytical Framework for Nuclear Nonproliferation. Vol. 8. Albuquerque, New México: Sandia National Laboratories; 2011. pp. 1-21

[12] Tenza A, Pérez I, Martínez-Fernández J, Giménez A. Understanding the decline and resilience loss of a long-lived social-ecological system: Insights from system dynamics. Ecology and Society. 2017;**22**(2):15

[13] Barrett C, Beckman R, Channakeshava K, Huang F, Kumar VA, Marathe A, et al. Cascading failures in multiple infrastructures: From transportation to communication network. In: 2010 5th International Conference on Critical infrastructure (CRIS); IEEE; 2010. pp. 1-8

[14] Spiegler VL, Naim MM, Wikner J. A control engineering approach to the assessment of supply chain resilience. International Journal of Production Research. 2012;**50**(21):6162-6187

[15] Tokgoz BE, Gheorghe AV. Resilience quantification and its application to a residential building subject to hurricane winds. International Journal of Disaster Risk Science. 2013;**4**(3):105-114

[16] Ergu D, Kou G, Shi Y, Shi Y. Analytic network process in risk assessment and decision analysis. Computers and Operations Research. 2014;**42**:58-74

[17] Sun H, Cheng X, Dai M. Regional flood disaster resilience evaluation based on analytic network process: A

case study of the Chaohu Lake Basin, Anhui Province, China. Natural Hazards. 2016;**82**(1):39-58

[18] Wang YM, Luo Y, Hua Z. Aggregating preference rankings using OWA operator weights. Information Sciences. 2007;**177**(16):3356-3363

[19] Al-Yahyai S, Charabi Y, Gastli A, Al-Badi A. Wind farm land suitability indexing using multi-criteria analysis. Renewable Energy. 2012;**44**: 80-87

[20] Jalao ER, Wu T, Shunk D. A stochastic AHP decision making methodology for imprecise preferences. Information Sciences. 2014;**270**:192-203

[21] Hu J, Mehrotra S. Robust and stochastically weighted multiobjective optimization models and reformulations. Operations Research. 2012;**60**(4):936-953

[22] Conde E, MdlPR P. A linear optimization problem to derive relative weights using an interval judgement matrix. European Journal of Operational Research. 2010;**201**(2):537-544

[23] Yager RR, Alajlan N. Probability weighted means as surrogates for stochastic dominance in decision making. Knowledge-Based Systems. 2014;**66**:92-98

[24] Yamagata Y, Maruyama H. Urban Resilience. Springer; 2016

[25] Francis R, Bekera B. A metric and frameworks for resilience analysis of engineered and infrastructure systems. Reliability Engineering and System Safety. Switzerland; 2014;**121**: 90-103

[26] Thomas L, Vargas L. Decision Making with the Analytic Network Process. German: Springer; 2006

[27] MacKay D. An example inference task: Clustering. Information Theory, Inference, and Learning Algorithms. 2003;**20**:284-292

[28] Tibshirani EB. An Introduction to the Bootstrap. Monographs on Statistics and Applied Probability. New York, London: Chapman & Hall/CRC; 1993. p. 1

[29] Efron B et al. Second thoughts on the bootstrap. Statistical Science. 2003; **18**(2):135-140

[30] Greco R, Di Nardo A, Santonastaso G. Resilience and entropy as indices of robustness of water distribution networks. Journal of Hydroinformatics. 2012;**14**(3):761-771

[31] Zurlini G, Li BL, Zaccarelli N, Petrosillo I. Spectral entropy, ecological resilience, and adaptive capacity for understanding, evaluating, and managing ecosystem stability and change. Global Change Biology. 2015; **21**(4):1377-1378

[32] Comfort LK, Siciliano MD, Okada A. Resilience, entropy, and efficiency in crisis management: The January 12, 2010, Haiti earthquake. Risk, Hazards & Crisis in Public Policy. 2011;**2**(3):1-25

[33] Comfort LK, Oh N, Ertan G. The dynamics of disaster recovery: Resilience and entropy in hurricane response systems 2005–2008. Public Organization Review. 2009;**9**(4):309

[34] Tamvakis P, Xenidis Y. Comparative evaluation of resilience quantification methods for infrastructure systems. Procedia-Social and Behavioral Sciences. 2013;**74**: 339-348

[35] Bailey KD. Social Entropy Theory. Albany, New York: State University of New York Press; 1990

[36] Bailey KD. Entropy systems theory. In: Systems Science and Cybernetics, in Encyclopedia of Life Support Systems (EOLSS), Developed under the Auspices of the UNESCO. Oxford, UK: Eolss Publishers; 2001. [http://www.eolss.net]

[37] Shannon CE. A mathematical theory of communication. ACM SIGMOBILE Mobile Computing and Communications Review. 2001;**5**(1): 3-55

[38] Vineis P, Chan Q, Khan A. Climate change impacts on water salinity and health. Journal of Epidemiology and Global Health. 2011;**1**(1):5-10

[39] Hongyi W. Shanghai drinking water hit by salt tide. ChinaDaily: Society. 2014. http://usa.chinadaily.com.cn/china/2014-02/24/content_17300378.htm [February 24, 2014]

[40] Jiayun K. Salt Tide Worst on Record, Hits 2m. ShanghaiDaily.com: Metro. 2014. http://www.shanghaidaily.com/Metro/environment/Salt-tide-worst-on-record-hits-2m/shdaily.shtml [February 24, 2014]

[41] Li Y, Chen C, Zhang X, Dai J, Xu H. Impact of salt-tide on urban water supply and its countermeasure. Water Resources and Hydropower Engineering. 2008;**10**:005

[42] Miao R, Huang W, Pei D, Gu X, Li Z, Zhang J, et al. Research on lease and sale of electric vehicles based on value engineering. International Journal of Production Research. 2016;**54**(18): 5361-5380

[43] Zheng G. 16th CMA Order. China Meteorological Administration. 2007. http://www.cma.gov.cn/2011zwxx/2011zflfg/2011zgfxwj/201208/t20120803_180761.html [June 11, 2007]

[44] Hammouda K, Karray F. A Comparative Study of Data Clustering Techniques. Ontario, Canada: University of Waterloo; 2000

[45] Chaudhari B, Parikh M. A comparative study of clustering algorithms using WEKA tools. International Journal of Application or Innovation in Engineering & Management (IJAIEM). 2012;**1**(2): 154-158

[46] Ghosh S, Dubey SK. Comparative analysis of k-means and fuzzy c-means algorithms. International Journal of Advanced Computer Science and Applications. 2013;**4**(4):35-39

Chapter 4

The Study of Risk Assessment of Soil Liquefaction on Land Development and Utilization by GIS in Taiwan

Lien-Kwei Chien, Jing-Ping Wu and Wen-Chien Tseng

Abstract

The issue of soil liquefaction has been investigated widely in the past 50 years. However, there is no an integrated method have been considered for the factors between regions' vulnerability of soil liquefaction and resilience capacity to perform the risk assessment of soil liquefaction hazard. This study selects Yunlin and Chiayi County as a demonstration area, and uses Model Builder of geo-processing models to connect multiple analysis processes, the liquefaction risk distribution in Yun-Chia Plain's area is carried out in 100 × 100 m grid map scale by GIS. The study results could provide the reference of land development and management in Taiwan.

Keywords: soil liquefaction, hazard risk assessment, pareto ranking method, land utilization management, GIS

1. Introduction

For global environmental changing, the change of land use cause severe and straightforward impact [1]. The form of land use is determined by several of complex factors, such as politics, economics, society, culture, and natural environment. However, if the change of land use is based on those complex factors, it will easily bring huge impact to the environment [2].

Reference [3] indicated that in hazard-prone areas worldwide, cyclones, flooding, landslides, earthquakes, and volcanoes are the most common natural hazards. The report showed that 73% of Taiwan's citizens are exposed in the three or more hazards; 99% of people are exposed in two or more hazards. This analysis reveals is the extent to which, at global and regional scales, there is substantial overlap between different types of hazards and population concentrations. The result shows that large percentages of the population in Taiwan reside in hazard-prone areas.

Reference [4] raises a new concept about three essential factors to cause a disaster such as Hazard, Exposure, and Vulnerability. In other words, risk management will rise to the top priority if we cannot predict or control the natural disaster. Mastering exposure and vulnerability is the first step. Reference [5] had pointed out that some of the researches that will consider Exposure as a part of Vulnerability

IntechOpen

analysis. For instance, United Nations Development Program (UNDP) presented a brand new concept to explain how the risk be assessed, in which risk = Hazard × Exposure × Vulnerability. Report from [6] shows how Vulnerability and Exposure can be decided and the potential disaster happened under climate change.

Kaohsiung, Taiwan Meinong earthquake measuring 6.6 on the Richter scale happened in February 6, 2015; and became one of the most harmful disaster, which caused serious damage to the society. The damages are restricted in causing building failure and also make tons of people in suffer. Central Geological Survey (CGS) of MOEA has also started to do large-scale national geological drilling survey to set up Taiwan's own soil liquefaction potential inquiry system.

As high speed growth of Taiwan's population and rapid economic development, land use is becoming more and more diversity. Disaster impact will be changed in accordance to dissimilar regions, environments, and soil types. In summary, the main principle of this study is to use the Geographic Information Systems (GIS) tool, Chiayi and Yunlin plain survey result done by National Land Surveying and Mapping Center, Ministry of the Interior (NLSC); and combine with hazard, vulnerability, and resilience maps to do the soil liquefaction risk mapping. Based on the soil liquefaction risk map, proper development, and utilization strategy of the region will be established. This study also could become a reference of a general land protection strategy and regional development licensing and risk assessment and management.

2. Research methodology

This study takes the risk of soil liquefaction in Chiayi and Yunlin plains, adopt those soil liquefaction potential index assessment which is commonly used in Taiwan and worldwide as the basic reference. United Nations Disaster Relief Organization (UNDRO) [7] proposed an operability definition for disaster risk: R(Risk) = H(Hazard) × V(Vulnerability), which is the most commonly used method of disaster risk assessment The study also uses this definition as a basis for assessing soil liquefaction risks. In order to fully realize the effect of land use, this study changes the past boundaries of risk, such as townships, towns, cities, and districts as the boundaries of risk allocation and takes the 100 × 100 m grid as the regional space unit.

2.1 Risk matrix

This study takes resilience as an indicator that represented how the region recovers from the disaster. Therefore, the risk assessment method used in this study is as Eq. (1) and the concept of risk matrix is shown in **Figure 1**.

$$\text{Risk} = \text{H(hazard)} \times \text{V(vulnerability)} \times \text{R(resilience)} \tag{1}$$

This study considered soil liquefaction as a cause of disaster, which is defined as hazard, the different phases of Taiwan society developments are named as Vulnerability and how Taiwan's response to upcoming disaster as Resilience. Taiwan soil liquefaction risk assessment will be done by combining hazard, vulnerability, and resilience into risk matrix.

In this study, CGS's drilling investigation result is selected as our database. Therefore, the assessment methodology used in this study is the same as CGS, which is called NJRA method [8]. This study adopts the soil liquefaction potential

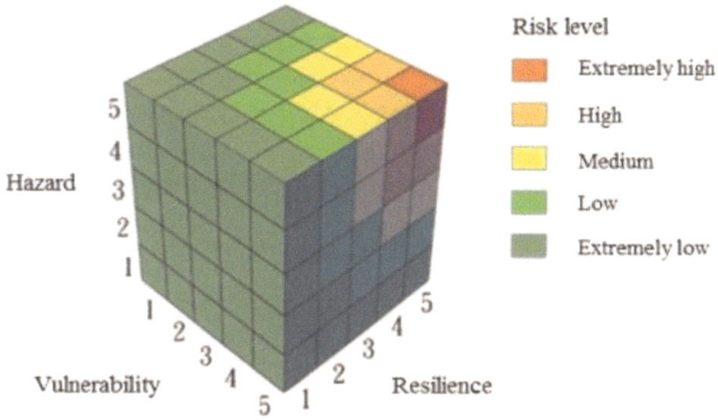

Figure 1.
The concept of the three-dimensional risk matrix.

mentioned in NJRA method, and considers different earthquake force designs as our hazard level grading. By comparing this study's result and CGS's result, people can figure out the difference between using normal soil liquefaction potential and our research's soil liquefaction potential that considers hazard, vulnerability, and resilience.

Second, vulnerability is based on the vulnerability indicators established by [9, 10]. To ensure that the database is correct, this study generally reviews the integrity of the indicators' data; ensures the indicators are selected appropriately and reflects the true development of the regions. Resilience indicators used in this study refer to [11, 12] for using the resilience indicators at all levels to consider the data integrity, the appropriateness of indicators which could showed the real condition of representative regions' development, and recovery. What is more, there is neither direct correlation nor the preference between the indicators. Therefore, the Pareto ranking (PR) method is used to standardize all the grading and abandon indicator weighing in order to establish the vulnerability and resilience.

This study seeks to simplify the mapping of potentially liquefiable areas by using Geographical Information Systems (GIS), and compared with the assessment method by the UNDRO to produce a map that could be used to evaluate potential damage in high risk areas of the county.

Cone penetration tests are used by the CGS in order to create GIS maps, but this is costly and not every county can afford to fund such studies on land. Geologic GIS layers are available due to the soil mapping of most counties by the CGS. Using the databank of social-economy and humanities in the county, along with other available data layers to narrow down the liquefaction risk of areas within the counties using ArcGIS and liquefaction criteria, the GIS layers for the Yun-Chia plain areas were modeled to select out the most liquefiable areas. These data layers were then combined to create a liquefaction risk map for Yunlin County and Chiayi County. This research can easily represent results straight forwardly.

2.2 Pareto ranking (PR) analysis

Before using PR analysis, those factors have to be standardized by Eq. (2). After getting a standardized grade, this study sums each region's vulnerability indicators total grade (grades of each factor vulnerability indicators are 1–5) and integrates those hazard indicators into six degrees. Hazard indicators are mean sea level rising, land subsidence, and storm surge flooding. And finally, use Pareto ranking (PR) analysis to evaluate general risk in coastal areas; and give each region a hazard risk level; the highest risk region gets level 9 and the lowest gets level 1, based on the principle to do PR analysis and evaluate each regional vulnerability.

Pareto ranking is a method for ordering cases on multiple criteria that has become popular in the context of genetic algorithms, where it is particularly valued because it often gives high rankings to those cases that only score heavily on one factor [13]. PR analysis is based on the principle of Pareto optimality. "Pareto optimality" is a formally defined concept used to determine when an allocation is optimal. An allocation is not Pareto optimal if there is an alternative allocation where improvements can be made to at least one participant's well-being without reducing any other participant's well-being. When no further Pareto improvements are possible, the allocation is a "Pareto optimum." A primary factor is selected from each vulnerability indicator. As shown in **Figure 2**, each point represents a grade of factor 1 and factor 2. If there is no point staying in the first quadrant, and there is no point ranked as high vulnerability degree. By taking same grade range region as a degree and following up the same step, this study can show each region's vulnerability distribution.

$$\text{z-score} = (X_i - M)/S \tag{2}$$

where Xi is the different indicators' data; M is the average grade, S is the standard deviation, $S = \sqrt{\left(\Sigma(X_i - M)^2/(N-1)\right)}$, and N is the number of data

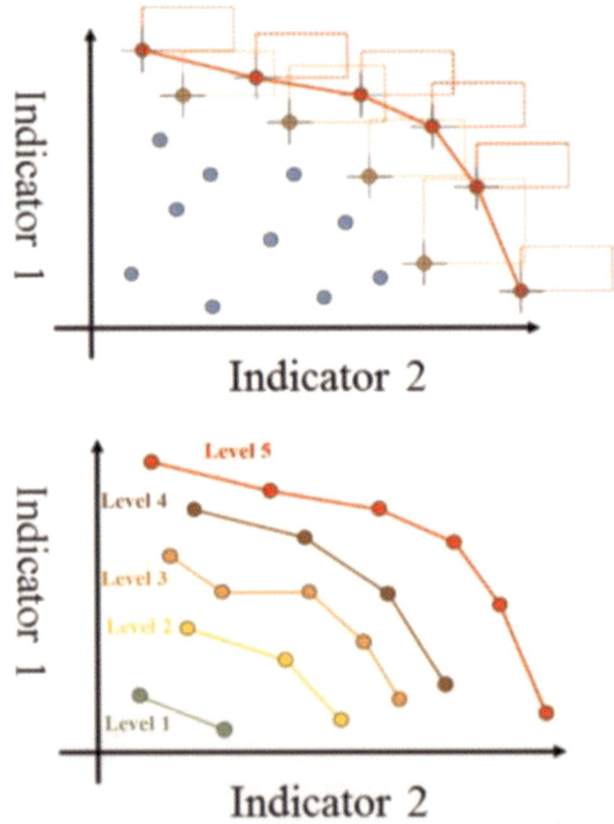

Figure 2.
Concept of PR analysis.

3. Analysis resources of database

3.1 General introduction of study area

According to the CGS disclosed geological drilling database, the Yunlin and Chiayi areas have relatively complete drilling data and complicated geographical environment. The area of environmental sensitive areas in Chiayi area of Yunlin is as high as 57%.

The strata along the western coast of Taiwan, especially the Chiayi area of Yunlin, are mostly modern alluvial strata. Because of their unconsolidated strata, the soil layers are mostly interbedded layers of sand and clay. In the past, soil liquefaction took place in sites containing particles with a diameter of about 0.01-cm loose fine sand-based and high groundwater level characteristics of the soil. Therefore, this study focuses on the Yushe River alluvial fan plain in Yunlin County, Chiayi County in the Chia-Nan Plain, the grid scale is 100 × 100 m, as shown in **Figure 3**.

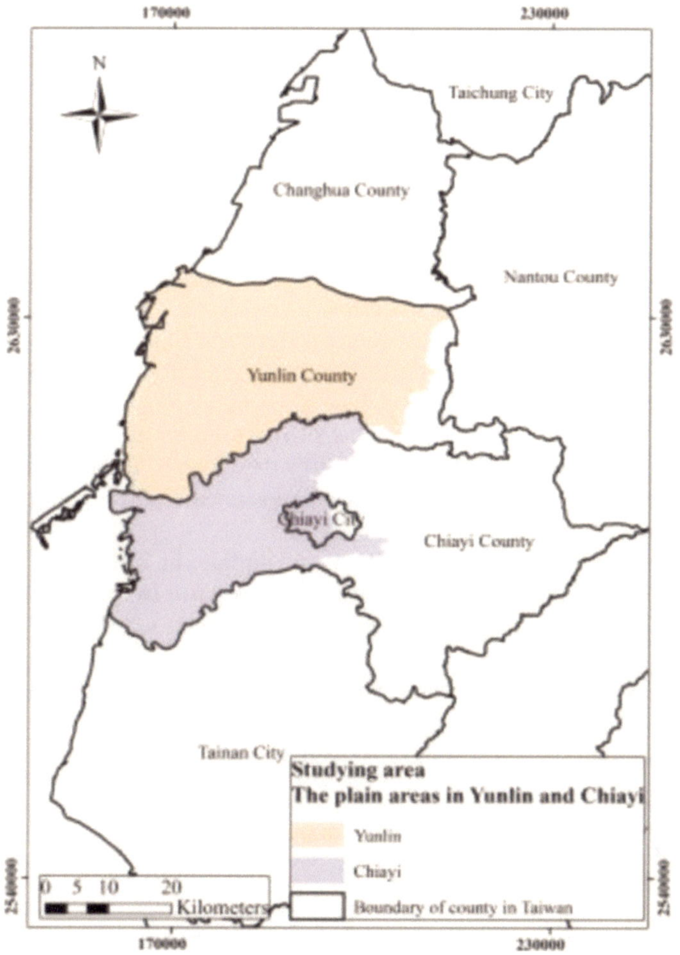

Figure 3.
Study area—the plate area in Yunlin and Chiayi.

3.2 Drilling data selection

According to the data of soil parameters for soil liquefaction assessment according to the NJRA method, there are eight items of data, including the Peak ground acceleration (PGA), fines content (FC), clay content (Pc), 10% particle diameter (D_{10}), average particle diameter (D_{50}), SPT-N value, depth of groundwater level (D_w), soil depth (D), and other parameters, which can be used as parameters that affect the liquefaction judgment as follows.

According to [14], three seismic forces of different peak ground acceleration (PGA) are designed in this study:

(I) Small and medium-sized earthquakes: the regression period of 30 years, the 50-year surpassing probability is about 80%, 0.067 g.

(II) Earthquake design: in the 475 years of regression, the probability of surpassing in 50 years is about 10%, 0.28 g.

(III) The biggest consideration earthquake: the regression period of 2500 years, 50 years beyond the probability of about 2%, 0.36 g.

Reference [15], based on the present soil data from the Niigata earthquake liquefaction area of Japan in 1964, concluded that liquefaction is less likely to occur when the soil fines content (FC) is greater than 35%; the selected liquefaction range: FC $\leq$ 35%.

Reference [16] showed that clay content (Pc) will exceed 20% and soil will not liquefy. Selected influence liquefaction range: Pc $<$ 20%.

Reference [17] proposed the average particle diameter D_{50} less than 10 mm and D_{10} less than 1 mm of sandy soil, liquefaction may occur; selected impact liquefaction range: $D_{50} \leq 10$ mm and $D_{10} \leq 1$ mm.

Reference [18] used the SPT-N value to determine that the sandy soil with N value of less than 10 for weak sites and N value less than 4 for clay. Selected influence liquefaction range: SPT-N $\leq$ 10.

Reference [19] pointed out that shallow soils tend to liquefy, studies have shown that liquefaction mostly occurs within a depth of 20 m of the soil, so a more conservative impact on liquefaction range is used: Groundwater depth (D_w) $<$ 10 m.

According to a study by Japan's Niigata earthquake in 1964, Ref. [20] found that liquefaction did not occur when the casing pressure in the area was greater than 2 kg/cm^2. Therefore, we set the depth of liquefaction: Soil depth (D) $\leq$ 20 m.

According to the above criteria of soil parameter selection, this study selected Yun-Chia area as the research area, excluding the borehole with incomplete data. Based on the calculation of the soil liquefaction potential, it was 850 in Yunlin area and 880 in Chiayi area, as shown in **Figure 4**.

3.3 Vulnerability

In order to investigate how vulnerability of soil liquefaction impact the urban and rural areas development in Yun-Chia Plain, this study considers the vulnerability references, data integrity, and the indicators of representative vulnerable areas. There are four indicators that represent factors such as population

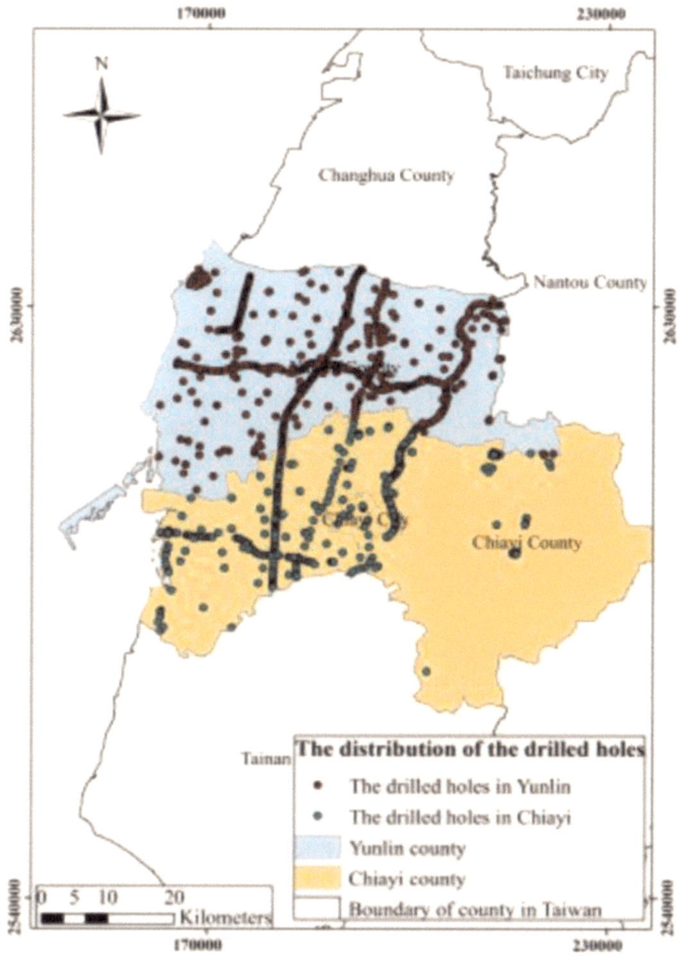

Figure 4.
The distribution of drilled holes in Yunlin and Chiayi (drew by this study).

density, gross industrial output, environmentally sensitive areas, and land use, which, respectively, representing the population distribution, economic development, urban and rural environmental development, and other indicators. Vulnerability included four indicators from the land surveys conducted by the National Land Surveying and Mapping Center in 2015.

3.4 Resilience

In order to investigate the resilience indicator which can show the make Yun-Chia plain areas recover from disasters. This study considers the resilience references, data integrity and the indicators of representative fragility areas. There are five representative indicators such as human development index, county budget, insurance coverage rate, social welfare workers, and communal participation rate, which represent social development, community development, government resources, property protection, and social assistance, respectively.

4. Analysis results

4.1 Hazard analysis in Yun-Chia plain areas

In this study, soil liquefaction is adopted as hazard. Referring to the maximum PGA of the three different design earthquakes force aforementioned for the NJRA method to assess soil liquefaction potential, also considering the effect of depth presented by [21], anti-liquefaction coefficient of different ground layers and the depth weighting factor are combined to obtain liquefaction potential index P_L of each drilling.

Previously, soil liquefaction potentials and different return periods of earthquake forces were usually discussed separately. This research takes different soil liquefaction potential indices PL induced by three different return periods of earthquake forces to separate the level of hazard, such as high, medium, and low. The three different return periods of earthquake forces are 30, 475, and 2500 years (**Figure 5a–c**), respectively. The basis classification of hazard degree is shown in **Table 1**. Hazard distribution condition is shown in **Figure 5d**.

4.2 Vulnerability analysis in Yun-Chia plain areas

The definition of vulnerability in this study is "the degree to which the area may be damaged." Vulnerability indicator introduction and calculation method are shown below.

Population density is a commonly used quantitative indicator reflecting the density of population distribution, which is the ratio of the total number of inhabitants under the unit area, as shown in Eq. (3). The density of grid units can reflect the seriousness of the people casualties in disaster-hit areas.

$$\text{Population density} = \left(\frac{\textit{village population}}{\textit{village area}}\right) \tag{3}$$

Gross industrial output is critical to measure the economic status and regions development level. It is the total output value of various industries in a unit area, as shown in Eq. (4). The level of the total industrial output within a grid of units can reflect the degree of economic loss that may be caused when a disaster strikes the area.

$$\text{Grid industrial total output} = \sum\left(\frac{\textit{gross industrial output}}{\textit{total industrial area}} \times \textit{industrial area in grid}\right) \tag{4}$$

Environmentally sensitive areas are basically equipped with special biological value or highly vulnerable to environmental impacts due to improper development activities. Based on the database platform established by the Construction and Planning Agency, Ministry of the Interior (http://60.248.163.236/SEPortal/), this study sets first-grade and second-grade environmentally sensitive area as the most vulnerable area, and non-environmentally sensitive areas are considered as low vulnerability.

Natural Breaks (Jenks) is used to speed up by dividing different PR scores into five different levels of vulnerability. The classification is shown in **Table 2**. The research of the result shows that the most vulnerable level is the Yunlin Industrial and Commercial Zone, the south of Koch Township, the Dongshih Township of Chiayi County, the coastal areas of Budai Township, and the residential areas in the center of Chiayi City. Yun-Chia plain area's vulnerability distribution and grading chart is shown in **Figure 6**.

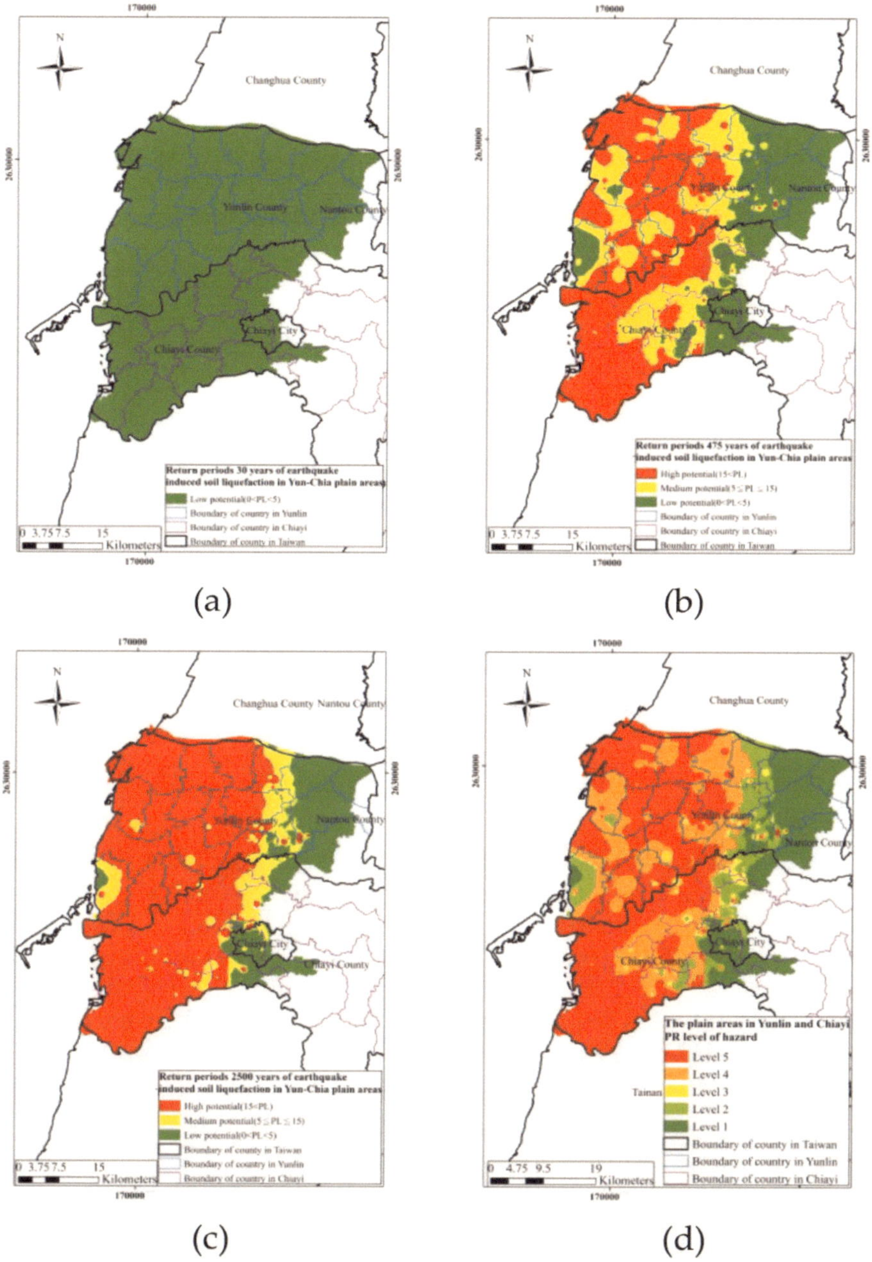

Figure 5.
Three different return periods of earthquake forces and Hazard distribution condition. (a) Soil liquefaction for earthquakes with a 30-year return period, (b) Soil liquefaction for earthquakes with a 475-year return period, (c) Soil liquefaction for earthquakes with a 2500-year return period, and (d) Hazard level induced by soil liquefaction.

4.3 Resilience analysis in Yun-Chia plain areas

The definition of resilience in this study is "the ability of the region to adapt frequent disasters, the ability of the plains to rebuild and improve after the disaster." Resilience indicator introduction and calculation method are shown below.

Due to the fact that this indicator is country scale, this study refers to the concept presented by [22], which took mortality, the proportion of higher

Level	The classification of hazard
5	The areas are high potential when the seismic design made by guideline happens ($15 < P_L$).
4	The areas are medium potential when the seismic design made by guideline happens ($5 \leq P_L \leq 15$). However, the areas are high potential when the maximum considered earthquake happens ($15 < P_L$).
3	The areas are medium potential when the seismic design made by guideline happens ($5 \leq P_L \leq 15$). In addition, the areas are medium potential when the maximum considered earthquake happens ($5 \leq P_L \leq 15$).
2	The areas are low potential when the seismic design made by guideline happens ($P_L < 5$). However, the areas are medium potential when the maximum considered earthquake happens ($5 \leq P_L \leq 15$).
1	The areas are low potential when the seismic design made by guideline happens ($P_L < 5$). In addition, the areas are low potential when the maximum considered earthquake happens ($P_L < 5$).

Table 1.
Classification of hazard.

	Classification	Level 1	Level 2	Level 3	Level 4	Level 5
PR score of distribution	Natural Breaks(Jenks)	0~2.14	2.14~4.06	4.06~7.21	7.21~9.80	9.80~14.18

Table 2.
The classification of vulnerability based on PR score.

education population, and the average individual income tax as alternatives to health, knowledge, and living standard. The higher the indicator, the better is its ability to cover from upcoming disasters. The data of this study are based on the report published by the Directorate-General of Budget, Accounting and Statistics, Executive Yuan, R.O.C. (Taiwan) in 2015. And its formula is calculated as shown in Eqs. (5)–(7). The Human development index (HDI) value is the geometric mean of the three basic indicators in Eq. (8).

$$\text{Life expectation index} = 1 - \left[\frac{(\textit{death rate} - 5)}{\textit{maximum mortality rate}}\right] \tag{5}$$

$$\text{Education index} = \frac{2}{3}\left(\frac{\textit{percentage of the population of higher education}}{45\%}\right) + \frac{1}{3}\left(\frac{\textit{literacy rate of population over } 15\,\textit{yrs}}{100\%}\right) \tag{6}$$

$$\text{Living standard index} = \frac{\left[\log\left(\frac{\textit{average household disposable income}}{\textit{US dollar}}\right) - \log(100)\right]}{[\log(75000) - \log(100)]} \tag{7}$$

$$\text{HDI} = \sqrt[3]{(\textit{life expectancy} \times \textit{education index} \times \textit{standard of living index})} \tag{8}$$

The counties and counties' financial budgets are related to the recovery potential after the counties and cities encountered disaster. The data of this study are based on the report published by Directorate-General of Budget, Accounting and Statistics, Executive Yuan, R.O.C. (Taiwan) in 2015. And its formula is calculated as shown in Eq. (9).

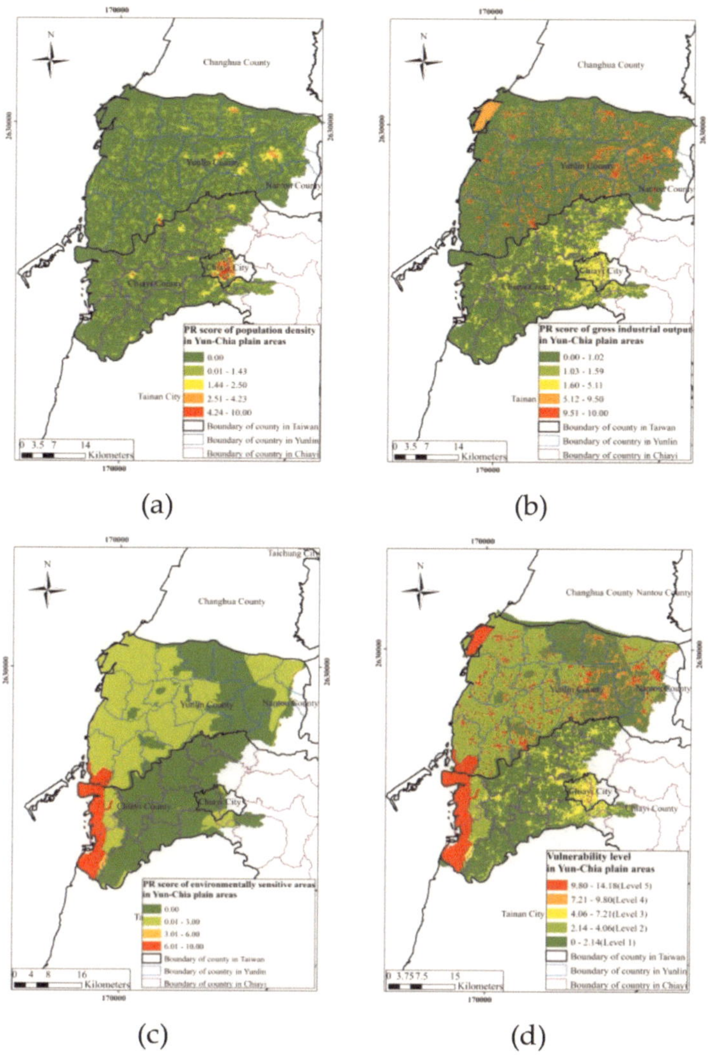

Figure 6.
Yun-Chia plain area's vulnerability distribution and grading chart. (a) Population density, (b) Gross industrial output, (c) Environmentally sensitive areas, and (d) Vulnerability distribution.

$$\text{County financial budget} = \left(\frac{county\ budget}{county\ population}\right) \times village\ population \tag{9}$$

Insurance coverage rate is the percentage given by the amount of earthquake insurance divided by the number of households in counties. The more insurance coverage rate, the more citizen can be recovered by the insurance after the disaster, and the higher the resilience.

Social welfare workers aim to prevent or relieve social problems by assisting individuals, families, groups, and communities in adapting their social functions, enhancing or restoring the energy of their social functions, and creating social conditions that achieve goals. The more staff members, the higher the resilience.

Communal participation rate reflects the local populace's income and preference for leisure. It is influenced by individual labor supply choices, which are affected by personal and social factors. Communal participation rate is the percentage given by the number of people willing to participate in community affairs divided by the number of people in a community. The higher the community participation rate, the higher the resilience.

Natural Breaks (Jenks) is used to speed up dividing different PR scores into five different levels of resilience. The classification is shown in **Table 3**. The research of the result shows that the highest resilience area is Maifeng Village, Yunlin County, HsinJei Village, Chiayi County, Ping Lin Lane, and Chiayi City District. Yun-Chia plain area's resilience distribution and grading chart is shown in **Figure 7**.

4.4 Risk assessment of soil liquefaction

This study sets the definition of disaster risk as a result of disasters caused by soil liquefaction such as structure failure, property loss, and even casualties. And hazard was defined as the earthquake-induced soil liquefaction; vulnerability is dined as population distribution, economic development, and environmental development; resilience is for social development, community development, government resources, property protection, and social assistance showing the recovery from disaster.

Taking the above definition as main principle, hazard, vulnerability and resilience are cross calculated and graded accordingly and divide the disaster risk levels into 1–5 points. The grading index standard of each factor given in this study is shown in **Table 4**. The risk analysis procedure showed how the combination among hazard, vulnerability, and resilience is shown in **Figure 8**.

Based on the report published in [23], dividing risk levels as five parts help the decision maker to make the optimized decision. This study followed the guide and divided risk into five categories: "extremely high," "high," "medium," "low," and "extremely low." The categories could easily show the risk when different area encountered the disaster. The risk of soil liquefaction distribution condition is shown as **Figure 9**. The risk assessment result assessed by UNDRO method (Risk = Hazard*Vulnerability) is shown in **Figure 10**.

By comparing the degree of vulnerability, resilience, and resilience used in this study, we can find that the risk analysis of joining resilience factor in this study is more effective than the risk analysis of two factors of degree of vulnerability and vulnerability in the past. When it comes to considering the resilience factor, the resilience of areas such as Mailiao Township, Lunbei Township, and Chiayi County of Yunlin City is better, and the degree of risk exposure in the event of a disaster is effectively reduced.

In this study, the soil liquefaction potential chart and the liquefaction risk chart obtained in 475 years of the return period of Yun-Chia plain and the soil liquefaction potential map in Yun-Chia area published by the CGS are analyzed and compared, as shown in **Figure 11**. It is found that the distribution of soil liquefaction potential obtained in 475 years (**Figure 5b**) of the study area of Yun-Chia Plain is similar to the distribution of soil liquefaction potential in Yun-Chia area published by the CGS. However, the risk analysis shows that most of the areas are high soil liquefaction potential area. But if we take the region's resilience into consideration, the risk level it faces in the event of soil liquefaction-induced disaster is not as severe as imagined.

	Classification	Level 1	Level 2	Level 3	Level 4	Level 5
PR score of distance	Natural breaks(Jenks)	0~5.24	5.24~6.10	6.10~7.33	7.33~9.36	9.36~13.41

Table 3.
Classification of resilience based on PR score.

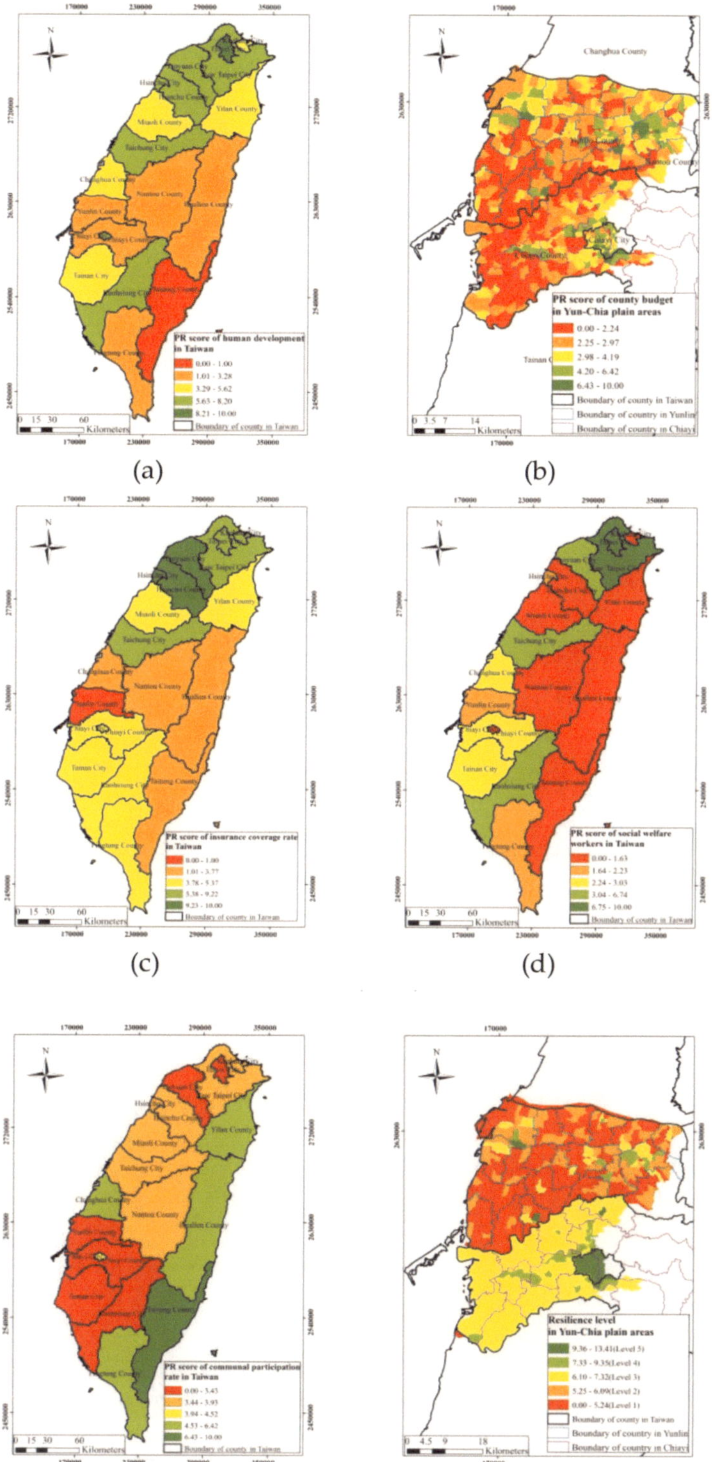

Figure 7.
Yun-Chia plain area's resilience distribution condition. (a) HDI distribution, (b) County budgetdistribution, (c) Insurance coverage rate distribution, (d) Social welfare workers distribution, (e) Communal participation rate distribution, and (f) Resilience distribution.

In formulating national strategy for disaster prevention, local governments can adjust and control the overall high soil liquefaction area and low potential in the event of soil liquefaction. The "Extremely High" and "High" risk region in this study should be the top priority improvements targets.

Indicator grading	Risk level 1	Risk level 2	Risk level 3	Risk level 4	Risk level 5
Hazard	1	2	3	4	5
Vulnerability	1	2	3	4	5
Resilience	5	4	3	2	1

Table 4.
Indicator grading and risk level.

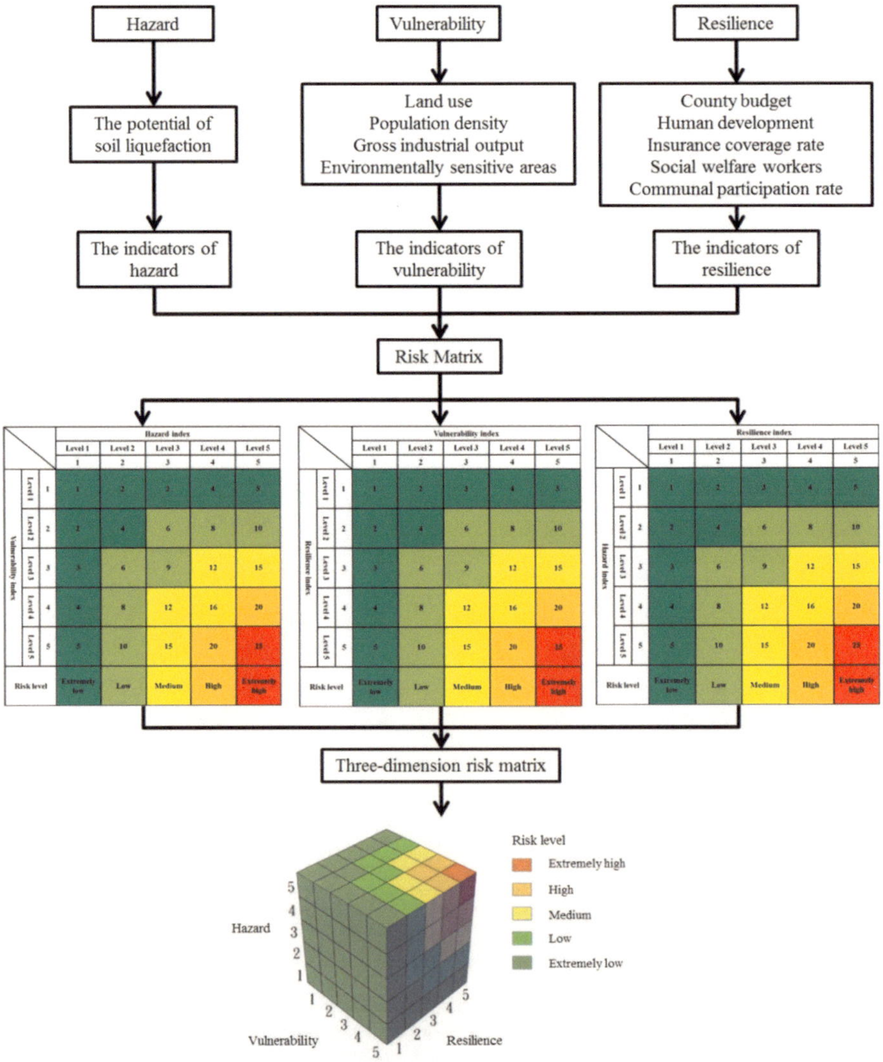

Figure 8.
Procedure of the risk level analysis.

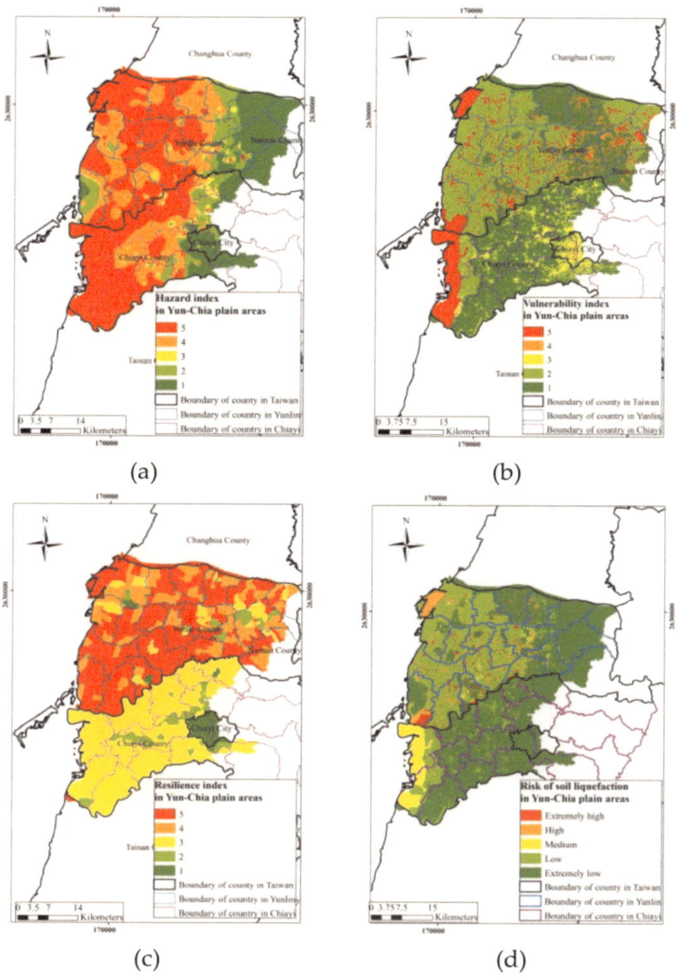

Figure 9.
Risk of soil liquefaction distribution condition. (a) Hazard index grading, (b) Vulnerability index grading, (c) Resilience index grading, and (d) Risk of soil liquefaction.

5. Conclusions

This study takes the risk of soil liquefaction in Chiayi and Yunlin Plains, adopt those soil liquefaction potential index assessment which commonly used in Taiwan and worldwide as the basic reference. This study also uses the definition proposed by UNDRO as a basis for assessing soil liquefaction risks but resilience was added into this study's risk assessment method, and considers risk as the combination of hazard, vulnerability, and resilience.

This research took different soil liquefaction potential index P_L induced by three different return periods of earthquake forces such as 30, 475, and 2500 years as Hazard. Vulnerability took four indicators representing factors such as population density, gross industrial output and environmentally sensitive areas, and land use. Resilience took social development, government resources, property protection, social assistance, and community development as the main points to show how the county recovers from the disaster.

According to the results of vulnerability analysis and resilience analysis, the vulnerability and resilience of Yunlin County are weaker than those of Chiayi

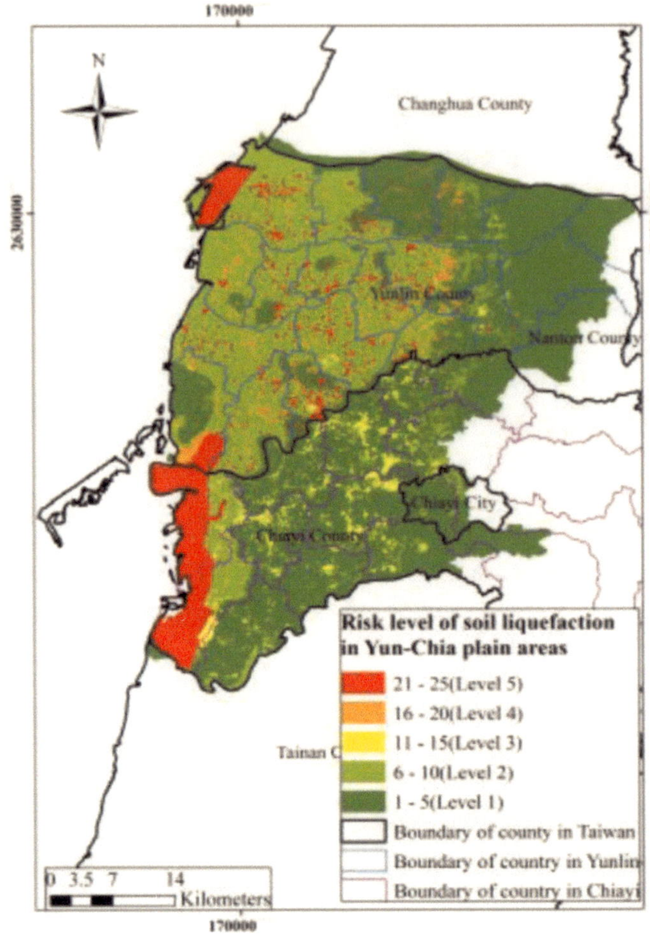

Figure 10.
The risk assessment result assessed by UNDRO method.

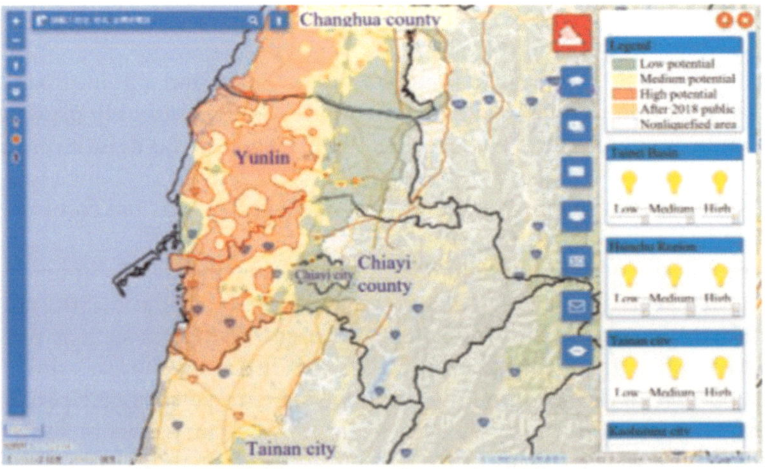

Figure 11.
Soil liquefaction potential map in Yun-Chia area published by the CGS.

County in the general areas, so in the event of a disaster, Yunlin County will likely face more serious losses in densely populated areas and high-output-value industries.

From the risk analysis results, we can see that in the face of the impact of soil liquefaction disasters, the population in Kouhu Township, Beigang Township, Tucu Town, Shui Lin Township, Dapi Township, Yuanchang Township, Dongsiang Township, and Baozhong Township of Yunlin County, was significantly lower than that of other large areas and low resilience led to these townships being over 10% at risk "extremely high" and "high," of which Kwuchu Township (26%) was the worst, followed by Beigang Town (16%) and Turku Town (14%).

This study discussed soil liquefaction from the perspective of risk. Compared with the CGS, "Soil liquefaction potential" can better reflect the relative risk among different regions and is more effective in concentrating decision makers in regional risk management.

Acknowledgements

The authors gratefully acknowledge the financial support of the Ministry of Science and Technology (No. MOST 106-2621-M-019-005-MY2) and Center of Excellence for the Ocean. Finally, the authors want to thank the reviewers for their very valuable comments.

Author details

Lien-Kwei Chien*, Jing-Ping Wu and Wen-Chien Tseng
Department of Harbor and River Engineering, and Center of Excellence for the Oceans, National Taiwan Ocean University, Keelung, Taiwan (R.O.C)

*Address all correspondence to: ntoulkchien@gmail.com

References

[1] Lambin EF, Turner BL, Geist HJ, Agbola SB, Angelsen A, Bruce JW, et al. The causes of land-use and land-cover change: Moving beyond the myths. Global Environmental Change. 2001; **11**(4):261-269

[2] Hasse JE, Lathrop RG. Land resource impact indicators of urban sprawl. Applied Geography. 2003;**23**(2–3): 159-175

[3] Dilley M, Chen RS, Deichmann U, Lerner-Lam AL, Arnold M, Agwe J, et al. Natural Disaster Hotspots: A Global Risk Analysis. Washington, DC: World Bank; 2005. 148 p

[4] Uitto JI. The geography of disaster vulnerability in megacities: A theoretical framework. Applied Geography. 1998; **18**(1):7-16

[5] Chiu SY. The analysis of coastal vulnerability in Taiwan area under climate change [Thesis]. Keelung: National Taiwan Ocean University; 2010 (in Chinese)

[6] Intergovernmental Panel on Climate Change. Summary for policymakers. In: Managing the Risks of Extreme Events and Disasters to Advance Climate Change Adaptation, A Special Report of Working Groups I and II of the Intergovernmental Panel on Climate Change. Kampala, Uganda: Cambridge University Press; 2012. pp. 1-19

[7] United Nations Disaster Relief Organization. Natural Disasters and Vulnerability Analysis: Report of Expert Group Meeting. Geneva, Switzerland: Office of United Nations Disaster Relief Co-ordinator; 1979

[8] Japan Road Association. Standards for Road Bridges and Commentary: V. Earthquake Resistant Design Edition. Tokyo, Japan: Maruzen Publishing Press; 1996. 318p (in Japanese)

[9] United Nations Environmental Program. Assessing Coastal Vulnerability: Developing a Global Index for Measuring Risk. Nairobi, Kenya: United Nations Environment Programme; 2005

[10] National Science and Technology Center for Disaster Reduction. Disasters Loss and Vulnerability Assessment of Society. NCDR 98-T11; 2010. 67p (in Chinese)

[11] Melissa P. Top-down assessment of disaster resilience—A conceptual framework using coping and adaptive capacities. International Journal of Disaster Risk Reduction. 2016;**19**:1-11. DOI: 10.1016/j.ijdrr.2016.07.005

[12] Hsiao MHH, Hsu KM. Social indicators of climate-related risk in Taiwan. Journal of City and Planning. 2015;**42**(1):59-86 (in Chinese)

[13] Rygel L, O'sullivan D, Yarnal B. A method for constructing a social vulnerability index: An application to hurricane storm surges in a developed country. Mitigation and Adaptation Strategies for Global Change. 2006; **11**(3):741-764. DOI: 10.1007/s11027-006-0265-6

[14] Ministry of the Interior. Seismic Design Specifications and Commentary of Buildings; 2011 (in Chinese)

[15] Ohsaki Y. Effects of sand compaction on liquefaction during the Tokachioki earthquake. Soils and Foundations. 1970;**10**(2):112-128

[16] Seed HB, Idriss IM, Arango I. Evaluation of liquefaction potential using field performance data. Journal of Geotechnical Engineering. 1983;**109**(3): 458-482

[17] Japan Road Association. Standards for Road Bridges and Commentary: V.

earthquake resistant design edition. Maruzen Publishing Press; 1990. 318p (in Japanese)

[18] Terzaghi K, Peck RB. Mekanika Tanah Dalam Praktek Rekayasa. Jakarta: Penerbit Erlangga; 1993

[19] Ishihara K, Yamazaki A. Analysis of wave-induced liquefaction in seabed deposits of sand. Soils and Foundations. 1984;**24**(3):85-100. DOI: 10.3208/sandf1972.24.3_85. Japanese Society of Soil Mechanics and Foundation Engineering

[20] Kishida H. Characteristics of liquefied sands during Mino-Owari, Tohnankai, and Fukui earthquakes. Soils and Foundations. 1969;**9**(1):75-92

[21] Iwasaki T, Arakawa T, Tokida K. Simplified procedures for assessing soil liquefaction during earthquakes. International Journal of Soil Dynamics and Earthquake Engineering. 1984;**3**(1):49-58

[22] Lu JC, Chen YM, Chang CH, Kuo YL. Evaluation of inundation risk under climate and environmental changes—An analytical framework for township level evaluation in Taiwan. In: 2009 Conference of Disaster Management of Taiwan. Taipei: Disaster Management Society of Taiwan; 2009 (in Chinese)

[23] University of Canterbury. AS/NZS ISO 31000:2009 Risk Management & Compliance Framework. New Zealand: UC Books; 2017

Section 3

Remote Sensing Data Analysis

Chapter 5

Modeling the Environment with Remote Sensing and GIS: Applied Case Studies from Diverse Locations of the United Arab Emirates (UAE)

Salem Issa and Nazmi Saleous

Abstract

Maintaining a healthy and sustainable environment is of paramount importance for human well-being and economic activity. Hence, environment protection, requiring continuous and reliable monitoring, has become a major mandate at various levels of government. There is a direct link between the availability of information about environmentally endangered areas and sound decision-making for effective and sustainable management. While remote sensing allows acquiring relevant information on a regular and repetitive basis even in areas where accessibility is limited, geographic information systems (GIS) provide unique tools for storing, processing, and integrating this information with other sources enabling the development of spatial models that help identify and characterize environmentally endangered areas. In this chapter, we will discuss how GIS-based modeling is applied for solving diverse environmental problems using case studies from United Arab Emirates.

Keywords: environment, remote sensing, GIS, modeling, urban growth

1. Introduction

Although deserts or arid lands typically do not have a large number of inhabitants, they are often the loci of economic and cultural activity. For example, the oil-producing nations of the Middle East are all found within a single arid region. Furthermore, deserts tend to be fragile ecosystems, requiring little in the way of perturbations in order to cause tremendous changes in the landscape [1, 2]. The size, remoteness, and harsh nature of many of the world's deserts make it difficult and expensive to map or monitor these landscapes or to aid planning for and management of renewable natural resources. The situation exacerbates in developing countries where lack of accurate maps and the need for rapid and relatively accurate mapping techniques are urgent; this is becoming challenging if we know the dimension of large-scale engineering projects being implemented, particularly in the wealthy Gulf States [3, 4].

Remote sensing and GIS are promising new-time and cost-effective techniques to image remote arid and hyperarid lands. With the use of ancillary field data and the calibration of remote sensing inputs, data integration within a GIS can enhance the extraction of information from satellite imagery and has led to a synergistic approach in spatial data handling and modeling [5–7], hence improving the accuracy of a variety of outputs [8, 9].

In this chapter, we will expose the power and benefit of integrating remote sensing and geographic information systems to model our environment through various case studies applied to the arid/hyper arid environment of the United Arab Emirates. Four case studies are introduced and discussed in Section 3 namely:

- Characterization of Al Ain city urban growth using multi-temporal remote sensing data and GIS [10].

- Assessing landfill locations for waste management for the city of Abu Dhabi using GIS [11].

- Mapping sand dune fields in Abu Dhabi Emirate over the period 1992–2013 [12].

- GIS-based wind farm site selection model offshore Abu Dhabi Emirate, UAE [13].

2. Research approach

Remote sensing and GIS are incorporated into environmental modeling for addressing environmental issues and problems. The core of this approach is to use the power embedded in these geospatial techniques to develop and implement a GIS project. Remote sensing here is treated as the science, technology, and techniques used to acquire the wanted data concerning the study area, processing those data, extracting relevant information about the studied area, and exporting the resulting file into a geodatabase.

A typical GIS project includes (1) stating the problem; (2) defining the study area; (3) acquiring, preparing, and automating the data; (4) processing the data; (5) building the geodatabase; (6) analyzing; and (7) visualizing, mapping, and report writing.

2.1 Stating the problem

The first step of implementing a GIS project is to state the problem and identify the objective of the analysis. The following questions need be addressed: What is the problem to solve? How is it solved now? Are there alternate ways to solve it using a GIS? What are the final products of the project—reports, working maps, and presentation-quality maps? Will the data be used for other purposes? What are the requirements for these? This step is important because the answers to these questions determine the scope of the project as well as how you implement the analysis.

2.2 Defining the study area

This step delineates a narrowed boundary of an area of interest. The information from *Step* 2.1 tells us about the proper location where the problem occurred and addresses the possible questions and answers under interest.

GIS possesses many and convenient ways for demarkation of a project's boundary. ERDAS IMAGINE® and ArcGIS®, worldwide used GIS software packages, allow users to work with geographic information data by inputting and manipulating map layers in a comprehensive manner. In this chapter, we use ERDAS Imagine 2014 and ArcGIS 10.6 for implementing all remote sensing and GIS processes.

2.3 Acquiring the data

Before locating and acquiring the needed data, a list of criteria should have been set to address the identified objectives of the problem to be solved in the study area.

Consider the following two real-world examples:

Example 1: Landfill Locations for Waste Management of the City of Abu Dhabi Using GIS (**Table 4**) and,

Example 2: GIS-based wind farm site selection model offshore Abu Dhabi Emirate, UAE (**Tables 7** and **8**).

Furthermore, the methodology needs to be analyzed to establish what kind of data is needed. The most important question that needs to be answered is Why do I need these data? If the data are truly needed, then this question is easily answered. If not, then the data are most likely not necessary to solve the problem.

To be able to work with data in GIS, you need to understand the nature and procedural steps of working with GIS data such as dataset formats, dataset attributes, dataset completeness, coordinate systems, and dataset sources (see **Tables 1** and **5**).

Data type	Primary/ secondary	Date	Resolution/ accuracy	Source
Landsat MSS	Primary	29 January 1972	57 m	UAEU
Landsat TM	Primary	28 august 1990	28.5 m	UAEU
Landsat ETM+	Primary	23 March 2000	15 and 30 m	UAEU
Aerial photos	Secondary	1976 and 1983	1:5000 and 1:50,000	Al Ain Town Planning and Surveying Sector
Al Ain land use map	Secondary	2000	30 m	UAEU
IKONOS	Secondary	2000 and 2006	1 m	(TPSS)
Master Plan of the Al Ain region	Secondary	1986–2000 and 2000–2015	—	(TPSS)
Al Ain administrative boundary map	Primary	2008	—	(TPSS)
Demographic data	Secondary	1989, 1995, 2001, and 2005	—	(TPSS) & UAEU

Table 1.
List of primary and secondary data used in the research.

2.4 Processing and preparing the data

Remote sensing data need to be prepared before being used for information extraction. This operation is made up of two main sub-processes: *pre-processing* and *processing*.

Pre-processing: involves data restoration which means data correction. This involves radiometric, atmospheric, and geometric correction and map projection.

Processing: involves data enhancement, data classification, data validation, and data export to GIS format.

In a GIS project, data processed and exported from remote sensing will serve as one input into the GIS database. GIS has a database management system component to support the proper management of both spatial and attribute data. It also enables convenient linking and relating of various data records by their locations on a common coordinate system. Some common tasks should be executed during the data processing and preparing step; these are as follows:

Re-defining and re-projecting data: The purpose is to define or/and to convert a particular layer of data from one coordinate system to another. Working with GIS involves more than one GIS layer; therefore, acquired datasets may contain different projections. Different data projections lead to distortion of data and inaccuracy in the analysis.

Conversion between raster and vector data models: File formats can also be varied in the forms of raster (*for example*, data derived from the remote sensing process) or vector (shapefile or feature class). Feature classes and shapefiles usually come embedded with attribute data, which allow the user to easily select and manipulate the information of interest. Therefore, converting a raster file to vector enables the user to intersect other available vector data.

Reclassification: To perform certain analysis in most cases, data need to be reclassified beforehand. Reclassification is a local operation that performs raster data analysis on a point-by-point or cell-by-cell basis. Reclassification, also commonly referred to as recoding, will reduce the number of classes you are using in the analysis, thus facilitating the analysis process and resulting in more accurate results. There are different reclassification methods such as binary masking, classification reduction, classification ranking, and changing measurement scales.

Data querying: Data querying in GIS involves both query by attribute and query by location. Both use certain conditions that apply to either the spatial or the non-spatial component of the analyzed data. The purpose is to extract desired features based on their location, attributes, or both for analysis. This can be done through conditional statement imposed in location or/and attribute data table to select only specific information of interest.

Data export: To make a temporary layer permanent in a current geodatabase, data resulted from steps such as that of the above need to be exported and saved in a working geodatabase or a current working space for further or future work.

2.5 Building the geodatabase

Creating the database for a GIS project will involve assembling the existing data, reviewing it, and then preparing the data for analysis. Some of the data will be usable as such; other layers will need additional processing. Sometimes you need to extract data from a possibly larger original source file. Reduction of the size of datasets and their consolidation accelerate the ensuing data processing and management. Typically, data acquired may exist in various forms and shapes, e.g. different coordinate systems and file formats. It is necessary to prepare and consolidate all datasets into a commonly operable form. GIS has a database management system

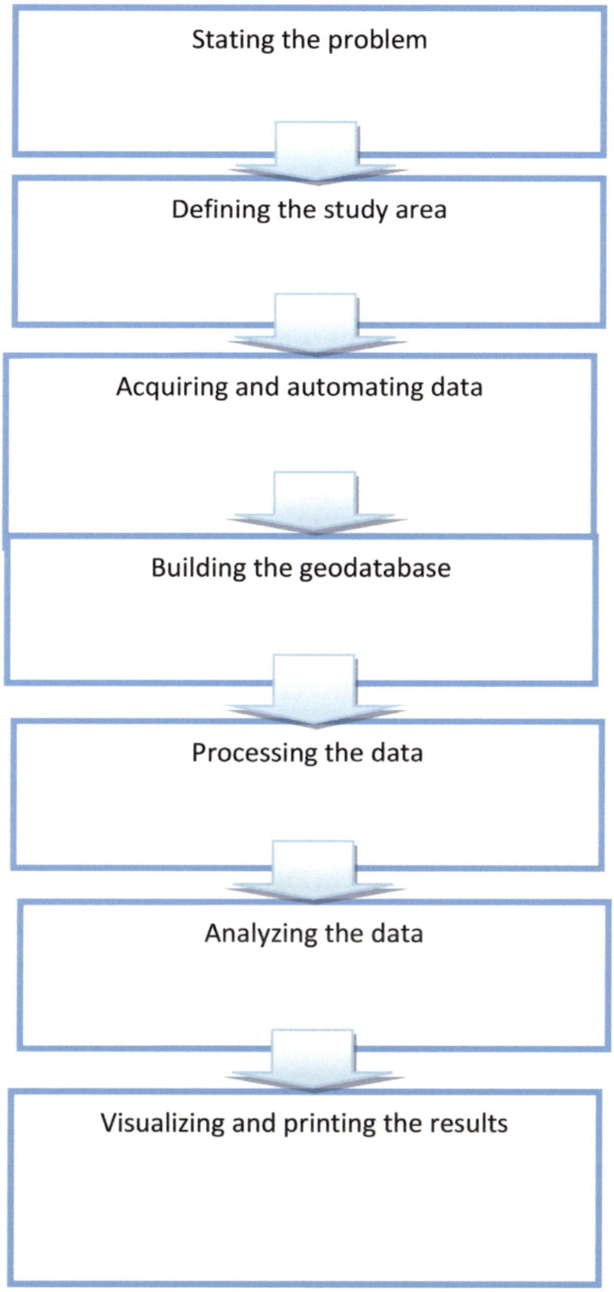

Figure 1.
Diagram of a typical GIS project: (1) stating the problem; (2) defining the study area; (3) acquiring, preparing, and automating the data; (4) processing the data; (5) building the geodatabase; (6) analyzing; and (7) visualizing, mapping, and report writing.

component to support the proper management of both spatial and attribute data. It also enables convenient linking and relating of various data records by their locations on a common coordinate system.

2.6 Analyzing the data

GIS analysis covers a wide variety of operations that you can do with a geographic information system. These range from simple display of features to complex, multistep analytical models. Perhaps the simplest form of GIS analysis is presenting the geographic distribution of data. This is conceptually the same as sticking pins in a wall map, a simple but powerful method of detecting patterns. A second type of GIS analysis is querying, or selecting from, the database. Queries let you identify and focus on a specific set of features. There are two types of GIS queries, *attribute* queries (find features based on their attributes) and *location* queries (find features based on where they are). A third type of GIS analysis is finding what is near a feature (buffering); a powerful function of GIS analysis is that the output of one procedure can be used in another. Here, the buffered zone can be used in an attribute query. A fourth type of GIS analysis is overlaying different layers of features. You can create new information when you overlay one set of features with another. There are several types of overlay operations, but all involve joining two existing sets of features into a single new set of features. Finally, all these techniques and many others are combined into a more complex GIS analysis, thus creating detailed models of the world to solve complicated problems. It is possible to repeat an analysis using slightly different parameters several times and compare the results. This can allow you to refine your analysis techniques.

2.7 Visualizing and printing the results

The last step in a GIS project is to present and communicate the results of your analysis. Your final product should effectively communicate your findings to your audience. The results of a GIS analysis can best be shown on a map. Nevertheless, they can also be disseminated through charts, reports, or videos and animated maps. You can print charts and reports separately, embed them in documents created by other applications, or place them on your map.

All the above steps are summarized in **Figure 1**.

3. Case studies

3.1 Characterization of Al Ain city urban growth using multi-temporal remote sensing data and GIS

The population of the UAE rose exponentially from around 86,000 in 1961 to more than 4 million in 2005. This has resulted in enormously rapid emplacement of a modern infrastructure, including an extensive highway and road networks, residential areas, shopping malls, golf courses, airports, and industrial facilities. The scale of such ambitious developments (often referred to as 'mega-projects') has been amazing and unmatched on a world scale.

In this case study, an attempt has been undertaken to map 'urban areas' in Al Ain city from large and medium-scale Landsat imageries in three different dates spanning the period 1972–2000 and to characterize the urban growth of the city using three different approaches: qualitative (using milestone change trajectories in the city), quantitative (using spatial metrics), and GIS overlay analysis.

3.1.1 Requirements for characterizing urban growth using remote sensing data

Capturing and analyzing the landscape change of the UAE have become key components to planners and policy makers in order to identify causes and assess the consequences of these changes on the future development of the society. Here rises the challenge of finding an effective way of measuring and documenting this change, sometimes very rapid, for a sustainable development that augments the people welfare while preserving the environment. Measurement and analysis of urban growth using remote sensing and geographic information system (GIS) techniques have seen very limited application examples in the UAE.

The study area is located between 55°28′ E to 55°53′ E longitudes and 24°03′ N to 24°22′ N latitudes (**Figure 2**). Al Ain is situated 150 km from Abu Dhabi capital city and 160 km from Dubai on the feet of Hafeet Mountain to the south and bordering Oman international boundaries to the east. The city is a perfect example of a small desert oasis with primitive society and limited resources to transform into a well-developed large city with an urban center hosting more than half a million inhabitants within a quarter of a century, making it an ideal example for urban growth studies using new remote sensing and GIS techniques in the region.

3.1.1.1 Datasets

A set of primary and secondary data is used in the research. Three Landsat satellite images from 1972, 1990, and 2000 (i. e., MSS1972, TM1990, and ETM + 2000) are processed and analyzed using ERDAS Imagine for the extraction of LULC classes in the three dates. Large-scale historical aerial photographs besides other ancillary data are used as reference data for accuracy assessment as well as in the geo-database building for further spatial analysis in GIS (**Table 1**). All images are atmospherically calibrated and geometrically rectified to a common Universal Transverse Mercator (UTM) coordinate system, zone 40, and the WGS84 Datum.

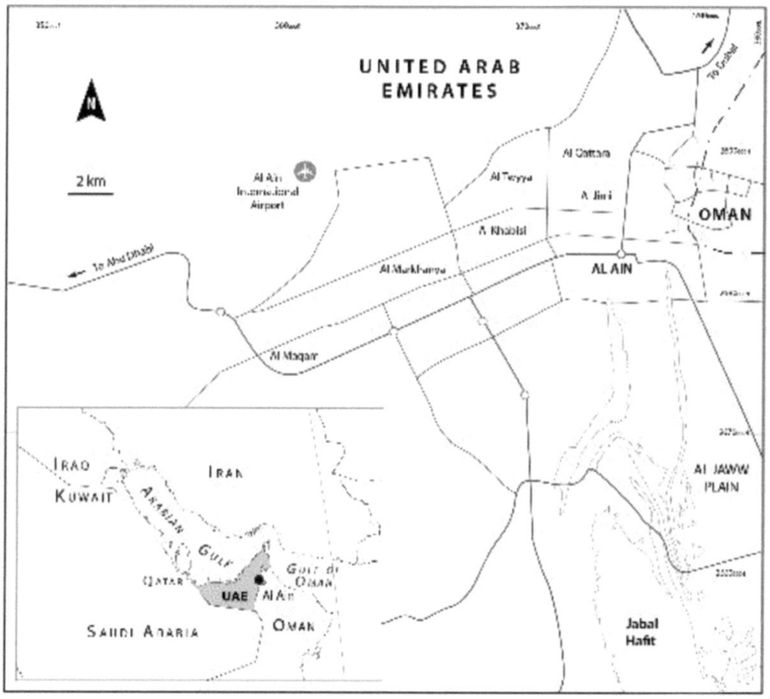

Figure 2.
Study area.

3.1.2 GIS project implementation for characterization of Al Ain urban growth

3.1.2.1 Classification schema

A hybrid of unsupervised and supervised classification schema is used. First unsupervised classification is carried out using the ISODATA algorithm. A number of iterations of 67, 80, and 60 for MSS, TM, and ETM+ are reached respectively with the convergence value at 0.990. The maximum likelihood algorithm with training sites carefully selected from the unsupervised classification results is used to run the supervised classification. Furthermore, the classified images were filtered using a 3 × 3 majority filter to remove speckles and to smooth the resulting images and decrease analytical errors.

The following are six classes representing most land cover types of the study area:

- Urban (built-up including roads and buildings)
- Vegetation (oases, farms, and parks)
- Sand and gravel (dark soils)
- Sand dunes (bright sand)
- Limestone
- Water bodies and shadow

3.1.2.2 Urban class extraction

Urban class is defined in this case as 'all manmade features including buildings, roads, and pavements in addition to vegetation covered areas such as oases, farms, parks, and farmed areas within the city boundary'. To extract urban areas, a bitmap is created to include urban and non-urban only.

3.1.2.3 Spatial metrics

The application of a number of spatial metrics is used to characterize the urban growth of the city. As such, the following spatial metrics are used:

- Land consumption rate (LCR)
- Land absorption coefficient (LAC)
- Annual urban growth rate (AGR)
- The percentage of built-up land (PLAND_U)

3.1.2.4 Change analysis techniques

The post-classification comparison approach is used based on comparing separately the produced classified LULC maps (1972, 1990, and 2000) in order to identify the change in the LULC classes and provide descriptive information about the nature of change that occurred in different dates. Spatial metrics and GIS overlay analysis are used to characterize changes in the urban area class. A total of eight instances of urban change trajectories are produced (**Figure 3**) [8].

3.1.2.5 Results

3.1.2.5.1 Urban class extraction

Urban areas were extracted using a semi-automatic method including manual editing of boundaries of certain classes based on authors' familiarity with the study area. A value of 1 was assigned to classes that fall in the urban category while a value of 0 was given to all other classes (**Table 2**).

3.1.2.5.2 Spatial metrics calculation

Results of the spatial metrics calculation are shown in **Table 3**.

3.1.2.5.3 Change detection between 1972, 1990, and 2000

Change detection analysis across 1972, 1990, and 2000 was conducted using the post-classification comparison method. The LULC classification results are presented in **Table 1**. The GIS overlay analysis was also applied on the LULC maps, which allowed the creation of 216 (=6^3) possible combinations of classes over the

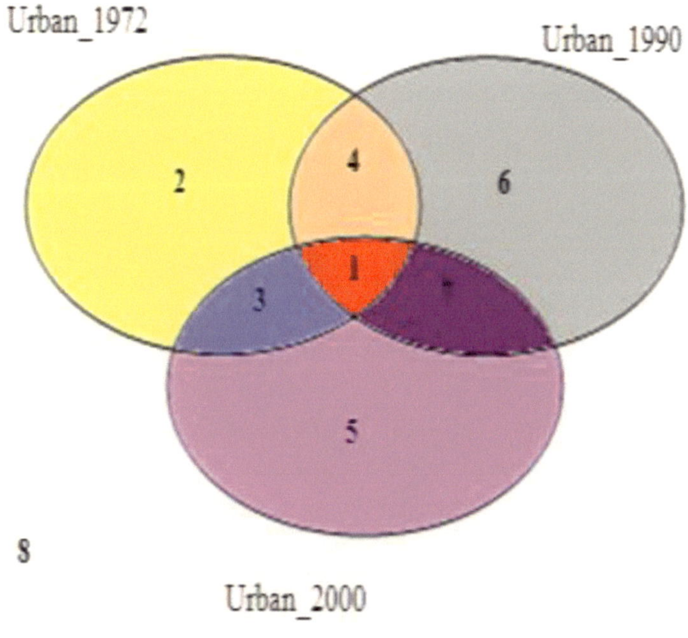

Figure 3.
Instances of urban change trajectories 72–2000.

LULC classes	1972		1990		2000	
	Area (ha)	Area (%)	Area (ha)	Area (%)	Area (ha)	Area (%)
Urban areas	4107.00	5.33	13,965.00	18.14	20,160.00	26.18
Others	72,893.00	94.67	63,035.00	81.86	56,840.00	73.82
Total	77,000	100.00	77,000	100.00	77,000	100.00

Table 2.
Areas (hectare) of urban areas.

study period, hence producing 216 different from-to-to change maps (not shown here). As our main concern was on the change characteristics of urban areas, a map was created using ArcGIS displaying the eight different instances of urban change trajectories throughout the study period (**Figure 4**).

In conclusion, it was found that neither Hafeet Mountain nor the sand dunes have ever formed a barrier to urban growth and will probably not do so in the future. Furthermore, it seems that in the short-term, the city will not witness urban expansion across the sand dunes for reasons explained above. However, it is thought that policy makers and planners will be forced to review their decisions to reverse the expansion from being horizontal, which was the case for the last forty years, to become vertical in order to minimize the necessary expenses of reclaiming more lands from sand dunes to urban areas, and to avoid more investments in kilometers of water pipes, electricity lines, and other infrastructures.

Year	LCR	PLAND_U (%)	Period	AGR (%)
1972	1.52	5.33	1972–1990	0.67
1990	0.43	18.14	1990–2000	0.73
2000	0.29	26.18		

Table 3.
Spatial metrics used for urban growth characterization.

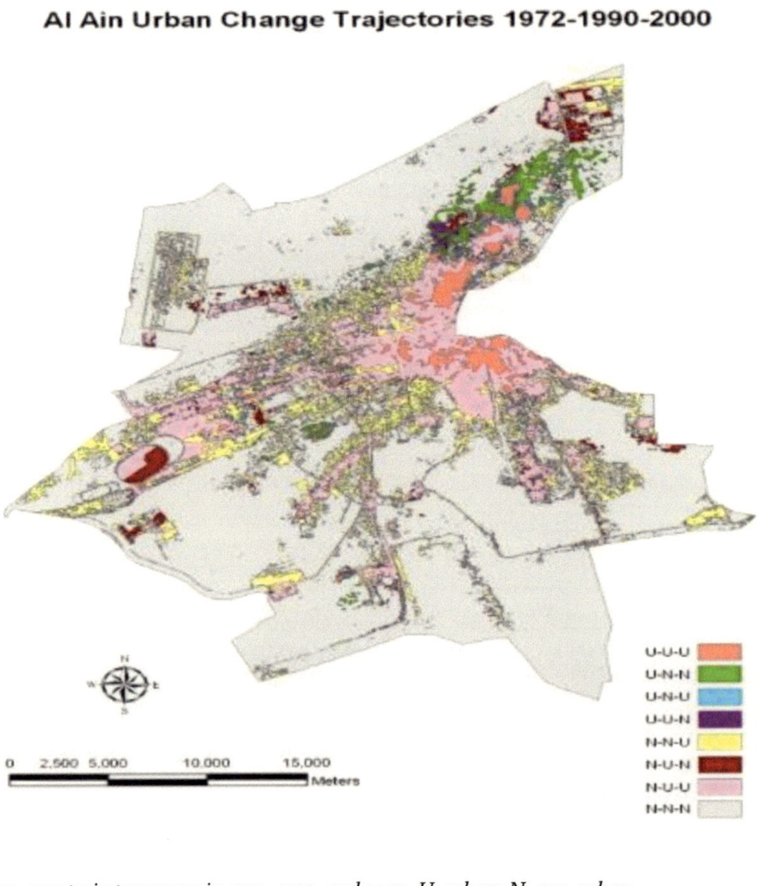

Figure 4.
Al Ain Urban area trajectory maps in 1972, 1990, and 2000. U, urban; N: non-urban.

3.2 Assessing landfill locations for waste management for the city of Abu Dhabi using GIS

The location of waste treatment and disposal facilities often has an impact on property values owing to noise, dust, pollution, unsightliness, and negative stigma. Hence, proper waste management practices need to be adopted to minimize the risks to human health and the environment.

3.2.1 Requirements for landfill location

A landfill must be situated and designed so as to meet the necessary conditions for preventing pollution of the soil, ground, and surface water as well as ensuring efficient collection of leachate. Similarly, a landfill site should be kept as far away from densely populated areas as possible, to reduce the impact of pollution on public health. At the same time, the landfill site should be placed as close as possible to existing roads to reduce costs of road development, transportation, and waste collection. Likewise, uneven or steep terrain is not appropriate for hosting landfills. As a result of an extensive literature review, local regulations, and expertise, a set of 19 criteria of an appropriate landfill site was identified and grouped into five parameters (**Table 4**).

3.2.2 GIS project implementation

The study area of the actual study is delimited by the administrative boundary of Abu Dhabi City Municipality. Datasets representing the set of 19 criteria identified in **Table 4** were collected from Abu Dhabi Municipality, the Environmental Agency of Abu Dhabi, Statistics Center of Abu Dhabi, and the National Center of Meteorology and Seismology. These datasets were clipped using the study area boundary, converted to a common spatial frame (UTM coordinate system, WGS84 datum, and Zone 40N), and added into a geodatabase for analysis.

The site selection process adopted a multi-criteria evaluation (MCE) approach that produces a suitability map from the criteria listed in **Table 4**. Given the varying importance of the parameters used in the selection process, they are assigned different weights as illustrated in **Table 4**. The overall suitability score (S) is then calculated based on the simple additive weighting (SAW) method as per Eq. (1).

$$S = \Sigma w_i \times x_i \tag{1}$$

where S is suitability, w_i is the relative weight of parameter i, and x_i is the rank (score) of parameter i. x_i is calculated as the sum of ranks of all the attributes belonging to parameter i and then normalized to the range 1–10 with 1 indicating the most favorable condition. The suitability value for each parameter was reclassified into four classes: Highly suitable (1–2), Suitable (3–5), Moderately suitable (6–7), and Unsuitable (8–10). The final suitability (S), calculated using Eq. (1) and reclassified into the four pre-defined classes, is shown in **Figure 5**. The final results indicate that 41% of the study area is considered as highly suitable for siting a landfill. Only 27% of the study area is unsuitable mostly due to proximity to restricted areas and oil fields.

The created suitability map can be used to select candidate locations for a new landfill site from available parcels that fall within the most suitable regions, such as the seven sites shown in **Figure 5**, prior to evaluating their environmental impact in order to help in the selection process. Furthermore, the model was used to examine and evaluate the existing “Al Dhafra Landfill” shown in **Figure 5**. It was concluded that this particular site falls within the unsuitable zone, owing to its proximity to oil concession areas.

Parameter	Weight	Layer name	Classification	Ranking
Environment	30%	Hydrology	<500 m	10
			>500 m	1
		Groundwater	<5 m	10
			5–15 m	5
			>15 m	1
		Geology	Sabkha, Eolian	1
			Other	10
		Slope	Slope: >15%	10
			5–15%	1
			<5%	5
		Shoreline	<500 m	10
			>500 m	1
		Vegetation	<200 m	10
			>200 m	1
Socio-economic criteria	30%	Roads	<1 km	10
			1–5 km	5
			5–10 km	1
			>10 km	10
		Built-up area	<1 km	10
			1–5 km	5
			5–10 km	1
			>10 km	10
		Land use	Residential < 2000 m	10
			>2000 m	1
			Industrial <200 m	10
			>200 m	1
			Other < 200 m	10
			>200 m	1
		Utilities	<500 m	10
			>500 m	1
		Population density	<2000 m from densely populated areas	10
			>2000 m from densely populated areas	1
Climatological criteria	10%	Wind speed	High	10
			Low	1
		Rainfall	High	10
			Low	1
		Temperature	High	1
			Low	10

Parameter	Weight	Layer name	Classification	Ranking
Restricted areas	20%	Airports	<3 km	10
			>3 km	1
		Military area	<3 km	10
			>3 km	1
		Oilfields	<3 km	10
			>3 km	1
		Protected areas	<500 m	10
			>500 m	1
Political criteria	10%	Admin boundary	<1000 m	10
			>1000 m	1

Table 4.
Landfill site selection criteria with weights and ranks.

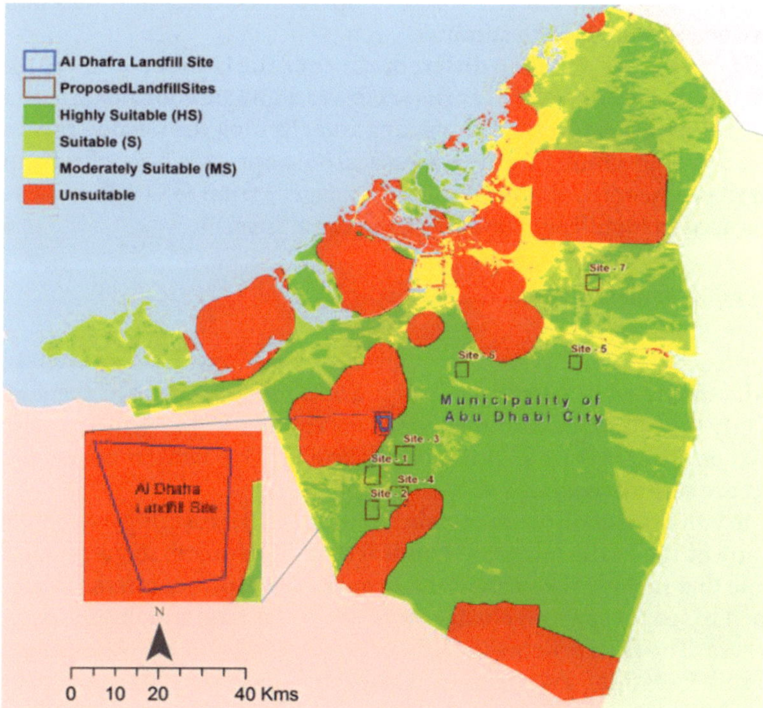

Figure 5.
The final landfill suitability map with the location of existing Al Dhafra landfill site and of potential sites for a new landfill (sites 1–7).

3.3 Mapping sand dune fields in Abu Dhabi Emirate over the period 1992–2013

The UAE has witnessed rapid economic growth since the discovery of oil in 1968. As a result, major urbanization and farming projects have been undertaken throughout the country including in the heart of the sand sea. Sand dunes and their movements present a serious threat to the country's urban centers as well as its infrastructure. This study focuses on the mapping of sand dune fields and assessing their changes over the period 1992–2013 in the UAE using Landsat imagery.

Availability of similar spatial and spectral characteristics of the Landsat TM and ETM+ sensors during the study period ensured the provision of consistent reliable imagery used in the study.

3.3.1 Requirements for mapping sand dune fields

Optical sensors have proven very useful in detecting sand over large areas. In particular, the use of discrete bands of multispectral sensors enables reliably to distinguish between sand and other land covers. Since the ultimate goal is to study changes in the sand cover extent, it is important that imagery used to map each representative period be acquired at anniversary dates.

In this study, we develop an approach to detect and map sand fields using Landsat imagery collected in three different years: 1992, 2002, and 2013. This approach is used to create land cover and sand/non-sand maps over the study area for the three different dates. We assess their accuracy using higher resolution imagery and store them in vector format for use in change detection studies. The study area encompasses the whole Emirate of Abu Dhabi extending over 67,000 km^2. It is characterized by a harsh climate where temperatures reach 48°C and humidity ranges between 80 and 90% in summer.

Six Landsat scenes that fall in different zones of the UTM coordinate system were used. It is necessary to convert the scenes to a unique coordinate system prior to creating a mosaic covering the study area and clipping the needed section. To delineate sand using a multispectral classification approach, a set of training and validation sites is needed. Higher resolution imagery from SPOT, Rapid Eye, and IKONOS is used in identifying and selecting these sites.

3.3.2 GIS project implementation

The approach used in this study is summarized in **Figure 6**. The scenes are mosaicked, referenced to a common frame, and clipped for each of the study periods. Only the reflective bands available on TM and ETM+ sensor and their counterpart on OLI are stored for use in processing. The datasets collected for use in the project are summarized in **Table 5**.

The core component of the processing consists in performing a supervised classification of the multi-temporal Landsat data. As a precursor, a classification scheme that includes important land cover classes present in the study area is developed based on spectral clustering of input datasets and familiarity with the study area. The final scheme includes the following classes: water, vegetation, sand, wet soil, intertidal, and bedrock. Different configurations of the processing flow showed that the overall accuracy of the classification increased if the vegetation class was extracted first using the soil adjusted vegetation index (SAVI). One hundred and forty two training sites selected randomly across the remaining classes are then used in the supervised classification process.

The resultant classified maps were reclassified into a binary sand/no sand map, vectorized, and exported to GIS. **Figure 7** shows the resulting land cover and sand/no sand maps obtained for the 3 dates. Accuracy assessment of the land cover maps was performed using a set of 94 assessment sites selected with the help of higher resolution imagery. The results indicated that the overall accuracy of the classification was 87% for the 1992 map, 89% for the 2002 map, and 91% for the 2013 map. However, sand class alone was mapped with a higher accuracy for all 3 years. **Table 6** summarizes the size of each one of the classes for the 3 time

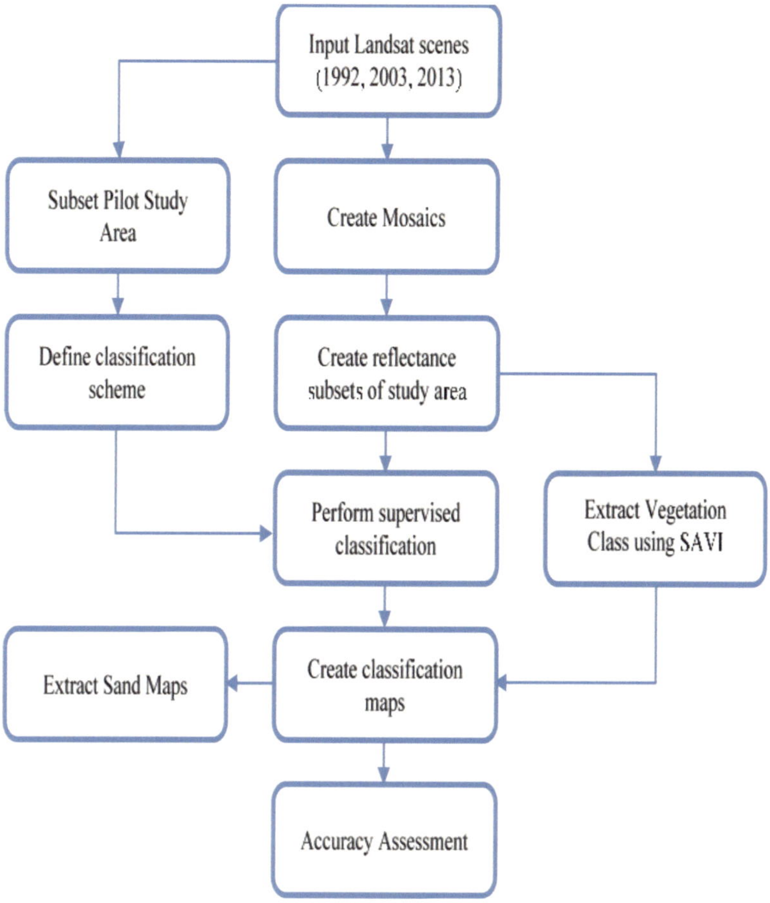

Figure 6.
Sand mapping methodology flowchart.

Data type	Date	Spatial resolution	Spectral characteristics	Purpose
Landsat 8 OLI	August 2013	30 m	Reflective bands	Classification
Landsat 7 ETM+	July 2002	30 m	Reflective bands	Classification
Landsat 4 TM	June 1992	30 m	Reflective bands	Classification
Rapid Eye	2013	5 m	RGB	Accuracy assessment
IKONOS	2003	1 m	Panchromatic	Accuracy assessment
SPOT panchromatic	1986	10 m	Panchromatic	Accuracy assessment
Abu Dhabi Emirate boundary shapefile	2013	—	Vector	Delimit study area

Table 5.
Datasets used in the study.

periods. It highlights the changes in the size of the sand class that can be attributed to different factors including sand encroachment, urban growth, establishment of farms, and dredging.

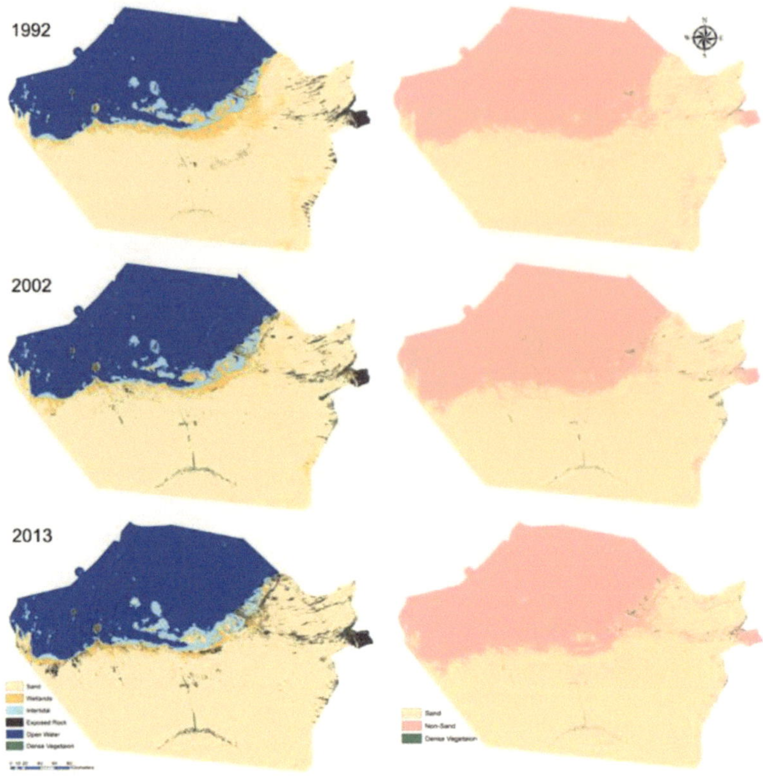

Figure 7.
1992, 2002, and 2013 land cover and sand maps.

Class	1992 (km^2)	2002 (km^2)	2013 (km^2)
Vegetation	147	518	440
Wet soil	5168	4341	2651
Intertidal	3069	2788	3175
Exposed bedrock	1874	1864	3821
Water	32,172	32,873	32,678
Non-sand land	10,258	9511	10,087
Sand	52,186	52,732	52,848

Table 6.
Size of the sand/non-sand classes for 1992, 2002, and 2013.

3.4 GIS-based wind farm site selection model offshore Abu Dhabi Emirate, UAE

The development of turbines that convert wind energy into electrical energy put wind in a good position as a source of alternative renewable energy. This study aims to assess the feasibility of establishing wind farms offshore the Emirate of Abu Dhabi, UAE and to identify favorable sites for such farms using a geographic information system (GIS).

3.4.1 Requirements for wind farm site location

Selecting the appropriate location for a wind farm is a key to its efficiency and success. Considering environmental, legal, and economic conditions, certain locations were found completely inadequate and should be excluded from the selection process, whereas others varied in their degree of suitability. **Table 7** lists the set of conditions that inhibit the siting of a wind farm and lead to the exclusion of areas with these conditions from the selection process. **Table 8** lists the set of criteria that affect the suitability of a location for a wind farm site.

3.4.2 GIS project implementation

Based on the set of criteria discussed in Section 3.4.1, needed datasets were collected, georeferenced to a common spatial frame, and ingested in the project's geodatabase. A GIS model is then built to create an exclusion mask from the input dataset (see **Figure 8**). Areas identified by the model are completely excluded from the selection process.

Areas that are not excluded by the first stage of the model are candidates for wind farms. Their suitability is evaluated using the criteria listed in **Table 8** using a weighted sum overlay approach whose inputs are derived from the wind speed and water depth layers.

The suitability of a location (S) is then calculated from the reclassified inputs using the weighted sum defined in Eq. (2).

$$S = w_{wind} \times \text{ranked wind speed} + w_{depth} \times \text{ranked water depth} \quad (2)$$

where w_{wind} is the weight given to the wind speed criterion and w_{depth} is that given to the water depth criterion.

Given the higher importance of wind in the suitability of a location for wind farming, we assigned a value of 2 to w_{wind} and 1 to w_{depth}. The suitability map resulting from running this model is presented in **Figure 9**. It indicates that only a small fraction of offshore Abu Dhabi Emirate is suitable for wind energy. A substantial

Parameter	Unsuitability condition
Water/land mask	Land
Submarine cables	Within 250 m
Oil and gas wells	Within 250 m
Oil and gas pipelines	Within 250 m
Bird conservation sites	Inside
Environmentally protected areas	Inside
Maritime navigation corridor	Inside

Table 7.
Criteria used to exclude areas from the selection process.

Parameter	Suitability condition
Wind speed	<4.5 m/s: unsuitable 4.5–5.4 m/s: moderately suitable >5.4 m/s: suitable
Water depth	0–10 m: moderately suitable 10–20 m: suitable >20 m: unsuitable

Table 8.
Criteria used to rank non-excluded areas.

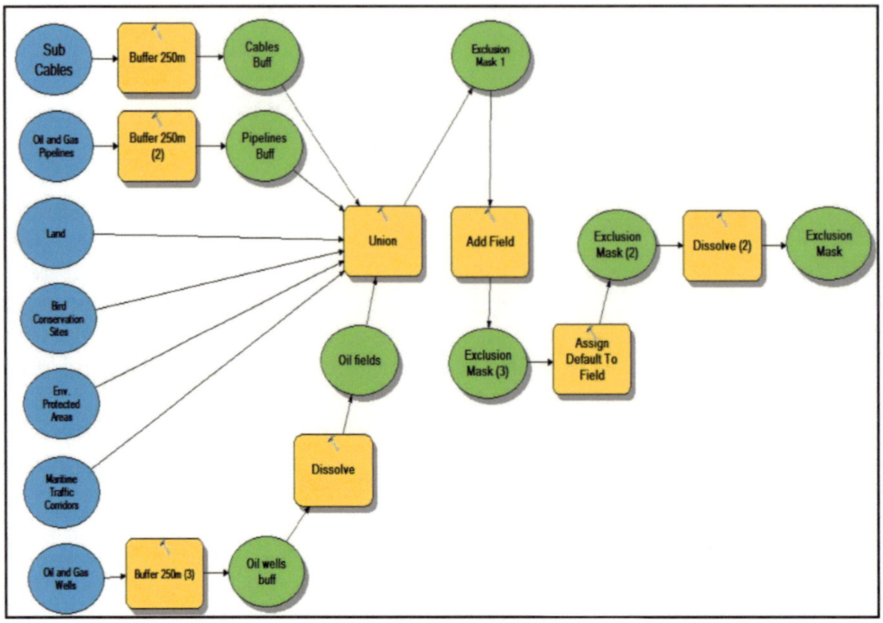

Figure 8.
Diagram of the area exclusion model.

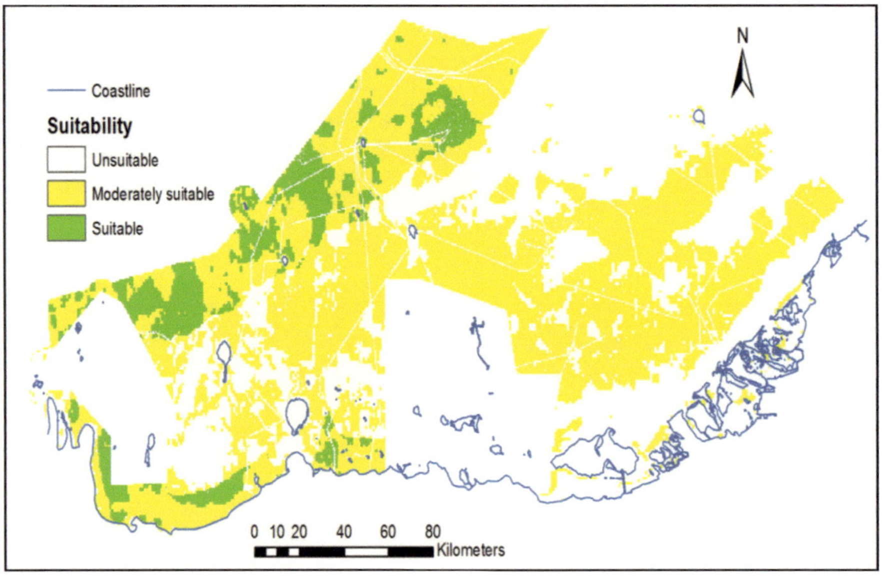

Figure 9.
Final suitability map for wind farm site selection offshore Abu Dhabi.

part of that area is considered moderately suitable. However, some suitable areas are close to mainland and to inhabited islands, such as Delma Island, and can be considered for wind farms to feed these areas.

4. Conclusions

In this chapter we highlighted the importance and outlined some methods of using satellite imagery integrated within GIS to address environmental issues facing human populations living in arid and hyperarid conditions. A systematic approach to use GIS-based models for solving environmental problems was detailed and illustrated in four different case studies to help readers gain better understanding of how GIS was deployed in environmental studies.

Environmental modeling starts with conceptualizing the real world into a mental model which is translated into a logical model implemented using a database management system and then transformed into a physical model to formulate a solution to the issue at hand. Processes and interpretations of the results are then implemented through the use of appropriate data and tools in a GIS environment. Expertise and knowledge to master these techniques will be best ascertained by hands-on practices.

Author details

Salem Issa* and Nazmi Saleous
Departments of Geology and Geography, United Arab Emirates University, Al Ain, UAE

*Address all correspondence to: salem.essa@uaeu.ac.ae

References

[1] Okin GS, Murray B, Schlesinger WH. Degradation of sandy arid shrubland environments: Observations, process modelling, and management implications. Journal of Arid Environments. 2001;**47**:123-144

[2] Starbuck MJ, Tamayo J. Monitoring vegetation change in Abu Dhabi Emirate from 1996 to 2000 and 2004 using Landsat satellite imagery. In: Second International Kuwait Conference; Kuwait; 2006. pp. 817-831

[3] Sohl T. Change analysis in the United Arab Emirates: An investigation of techniques. Photogrammetric Engineering and Remote Sensing. 1999;**65**(4):475-484

[4] Issa SM, Al Shuwaihi A. Analysis of LULC changes and urban expansion of the resort city of Al Ain /United Arab Emirates, using remote sensing and GIS. In: 2011 Proceedings of the 6th International Workshop on the Analysis of Multi-Temporal Remote Sensing Images (Multi-Temp); 12-14 July 2011; Trento. Italy; 2011. pp. 245-248. ISBN: 978-1-4577-1202-9. DOI: 10.1109/Multi-Temp.2011.6005094. http://ieeexplore.ieee.org/xpl/freeabsall.jsp?arnumber=6005094

[5] Jenesen JR. Remote sensing of the environment: An earth resource perspective. Upper Saddle River, NJ: Prentice-Hall, Inc; 2000

[6] Suga Y, Tsuru M, Tanaka S, Nakayama Y. Land resource database based on multiple satellites sensors and geographical information systems. Advances in Space Research. 1994;**14**:207-216

[7] Prost G. Remote sensing for geoscientists: Image analysis and integration. 3rd ed. Broken sound parkway NW, Boca Raton, FL: Taylor & Francis Group, CRC Press; 2014

[8] Salami AT. Vegetation dynamics on the fringes of lowland tropical rainforest of south-western Nigeria- an assessment of environmental change with air photos and Landsat. International Journal of Remote Sensing. 1999;**20**:1169-1181

[9] Smith M, Goodchild M, Longley P. Geospatial analysis. Comprehensive guide to principles techniques and software tools [Internet]. 2015. Available from: www.Spatialanalysisonline.com [Accessed: 2018-09-15]

[10] Issa S, Al Shuwaihli A. Characterization of Al Ain City Urban growth using multi-temporal remote sensing data and GIS. International Geoinformatics Research and Development Journal. 2012;**3**(1). http://www.igrdg.com/index.php/archive-of-issues/

[11] Issa SM, Saleous N, Al Shehhi B. Assessment landfill location for waste management for the city of Abu Dhabi using GIS. The Arab World geographer/Le Géographe du monde Arab. 2015;**17**(2):116-132

[12] Saleous N, Issa SM, Saeed R. Mapping sand dune fields in Abu Dhabi Emirate over the period of 1992-2013 using Landsat data. In: Pirasteh S, Li J, editors. Global Changes and Natural Disaster Management: Geo-information Technologies. Cham: Springer; 2017. pp. 101-112. DOI: 10.1007/978-3-319-51844-2_8

[13] Saleous N, Issa SM, Al Mazrouei J. GIS-based wind farm site selection offshore Abu Dhabi. In: Proceedings of the XXIII Congress of the International Society for Photogrammetry & Remote Sensing. The International Archives of the Photogrammetry, Remote Sensing and Spatial Information Sciences, Commission VI, WG VI/4. Volume XLI-B8, 2016. pp. 437-441. XXIII ISPRS Congress, 12-19 July 2016; Prague, Czech Republic. http://www.int-arch-photogramm-remote-sens-spatial-inf-sci.net/XLI-B8/437/2016/ doi:10.5194/isprs-archives-XLI-B8-437-2016

Chapter 6

GIS and Remote Sensing for Mangroves Mapping and Monitoring

Hamdan Omar, Muhamad Afizzul Misman and Samsudin Musa

Abstract

Malaysia is one of the few South East Asian counties with large tracts of mangroves. They provide ecosystem goods and services to the environment and the surroundings regarding shoreline stabilization, storm protection, water quality maintenance, micro-climate stabilization, recreation, tourism, fishing and supply of various forest products. Despite extensive distribution of the mangroves, threats posed by different land use activities are inevitable. Therefore, knowledge on mangroves distribution and change is importance for effective management and making protection policies. Although remote sensing (RS) and geographic information system (GIS) has been widely used to characterize and monitor mangroves change over a range of spatial and temporal scales, studies on mangroves change in Malaysia is lacking. Effective mangrove management is vital via acquiring knowledge on forest distribution and changes to establish protection policies. This chapter will elaborate technically how GIS and RS were utilized to identify, map, and monitor changes of mangroves ecosystem in Malaysia. It also highlights how GIS can enhance the current governance and regulations related to forestry in Malaysia.

Keywords: mangrove ecosystem, Landsat satellites, monitoring, deforestation, carbon emission

1. Introduction

Mangroves act as frontiers that protect the coastal land against destruction of ocean waves, tsunamis and storms. Mangroves also provide habitat for various aquatic life forms and function as natural filter, which improves the quality of water. Mangroves also play important roles as a significant carbon sink in coastal environment. It is interesting fact that despite only 0.05% of plant biomass stored in the ocean and coastal areas out of the total plant biomass on land, it can absorb a comparable amount of carbon every year. A study demonstrated that primary productivity in mangroves is higher than other types of forests. Biomass carbon in mangroves stands is among the highest in the tropics. Mangroves can store up to four times more carbon (C) as compared to other tropical forests around the world [1].

A mangroves ecosystem has an ability to absorb carbon dioxide (CO_2) and store carbon 40% more than the dry land forest ecosystem. Due to this ability, the total carbon deposited in a square kilometer of mangrove ecosystem is 50 times faster than those of the same area in a dryland tropical forest ecosystem. The absorbed CO_2 is stored not only in the plants, but in layers of soils underneath [2]. Therefore, mangroves are playing a crucial role in global carbon budgets and thus mitigating climate change.

However, despite being realized the importance of mangroves in the global carbon cycle and climate change, the extents of mangroves have inevitably declined since the last few decades. Unfortunately, the declines have been resulting mainly from human activities such as aquaculture expansion, coastal development, and over-harvesting [3]. Malaysia is one of the countries in South East Asia that has among the largest extents of mangroves. Despite its extensive distribution of mangrove ecosystem, this forest is inevitable from threats by various land use activities. The total area of mangrove forest was approximately 2% (650,000 ha) of the total land area in Malaysia in the 1990s [4].

However, the mangroves in Malaysia have been gradually diminishing, where the total area of mangrove forest has reduced to approximately 580,000 ha in the last decade [5]. Other reports indicated that the extent of mangrove areas in Malaysia is decreasing, from about 700,000 ha in 1975 to 572,000 ha in 2000 due to the intensive harvesting and natural wave actions [6, 7]. Globally, mangroves have also declined from 18.8 million ha to 15.6 million ha between years 1980 and 2005 [8]. Overall Asia was the largest net loss of mangroves since 1980, with about 1.9 million ha have loss, mainly due to conversion of mangrove forest to other land uses. However, there has been a slowdown in the annual rate of mangrove loss, from about 187,000 ha in the 1980s to 102,000 ha between 2000 and 2005. This reflects an increased awareness and an improved management system in mangroves ecosystem.

Major threats towards the mangroves that are triggered by human activities can generalized into six [9], which are (i) conversion to other uses, (ii) overharvesting, (iii) overfishing, (iv) pollution, (v) sedimentation and (vi) alteration of flow regimes. Direct conversion to other uses was identified as the major factor that changes the world's mangroves. This includes conversions to (i) urban and industrial areas, (ii) aquaculture, and (iii) agriculture. Additionally, natural phenomena such as coastal erosion, storm and lightning strikes are also the natural impacts that kill mangroves in Peninsular Malaysia, including the tragic tsunami on 24 December 2004.

Despite widespread concern and numerous case studies describing local issues and challenges, comprehensive information on the global extent of mangroves and trends of deforestation is largely lacking [10]. It is because determining the precise area of mangroves is not always easy. Measurement is affected by varying definitions of what constitutes mangroves; inclusion only on the basis of official recognition such as gazetted forest reserves; scattered or sparse areas considered too inconsequential for inclusion; and the accuracy of the returns made by the responsible authorities. Each of these can create uncertainty and produce significant variation depending on the timing and purpose of the assessment exercise.

Recently, RS satellites have been widely used for mangrove monitoring. They greatest reasons why is because the RS can (i) acquire information over large areas, (ii) produce repeated measurement over a place, and (iii) make full use of electromagnetic spectrum for quantitative and qualitative measurements over mangroves [11]. Satellites also provide information on spatial distribution and temporal changes of mangrove forests. When this information is gathered over decades, the mangrove monitoring over the large area will become possible. There are studies on the assessment of mangroves changes and identifying threats, for example in Terengganu [12], Selangor [13], and Peninsular Malaysia [14]. However, these studies are unable to represent the holistic conditions at national level. Therefore, this study was conducted to provide the information pertaining status of mangroves and changes that occurred since the last three decades.

2. The identification from remotely sensed data

2.1 The study area

The study area covers the entire mangroves ecosystem in Malaysia, which can be divided into two regions, which are Peninsular Malaysia and East Malaysia (i.e. Malay Borneo). Forests in these regions can be divided into three major types, which are inland dipterocarps (dryland), peat swamp, mangrove forests (wetlands). The mangrove forest is a unique ecosystem and the second largest wetland forest type after the peat swamp forest. Ecologically based on elevation the mangrove forest is located at the lowest elevation, which is equivalent to the sea level. The mangrove forest is generally found along sheltered coasts where it grows abundantly in saline soil and brackish water dominated mainly by trees from the *Rhizophoraceae* family. Mangroves are fringing the coastlines (up to 5 km landward) and major estuaries of the regions and they reside on wetlands ecosystem of not more than 20 m land altitude.

2.2 Satellite data

Images from Landsat-5 Thematic Mapper (TM), Landsat-7 Enhanced Thematic Mapper (ETM+), and Landsat-8 Operational Land Imager (OLI) satellite were used in this study. Images from three different epochs, which are 1990, 2000 and 2017 were acquired to conduct the work. For the respective years were utilized in this study. All images are available at https://earthexplorer.usgs.gov/ and were downloaded free of charge. At least 23 scenes of Landsat images were used for a single epoch (**Figure 1**). Therefore, to complete the series, the study has acquired at least 69 scenes of Landsat images, assuming that all images are free from cloud cover. However, cloud cover are presence on some of the images, hence, more than one scene of images over the same year were acquired to remove the clouds.

2.3 Production of seamless mosaic images

Cloud cover is inevitable on the images acquired by the satellites. However, cloud patching process can eliminate the cloud covers that appear on a single-date observation data. Images of particular scenes that were acquired on different dates were used for cloud patching process as shown in **Figure 2**. F_mask algorithm was used to perform this process [15, 16]. Seamless mosaics product (i.e. images without cloud covers and atmospherically corrected) were used as input for subsequent processes.

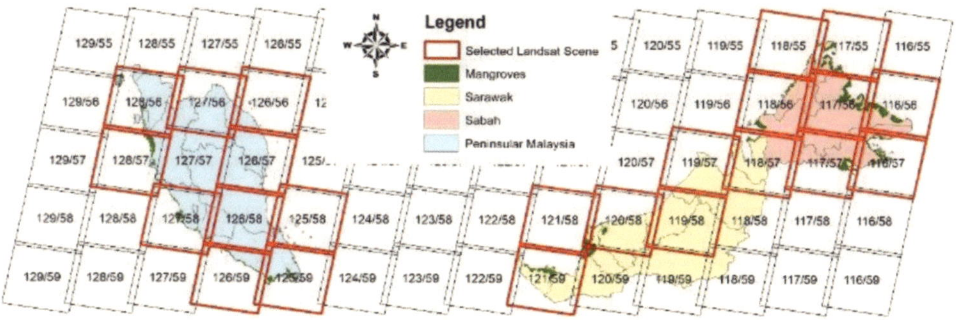

Figure 1.
Landsat scenes that were used for the classification. Numbers within the scene boundary indicates path/row ID of Landsat satellites.

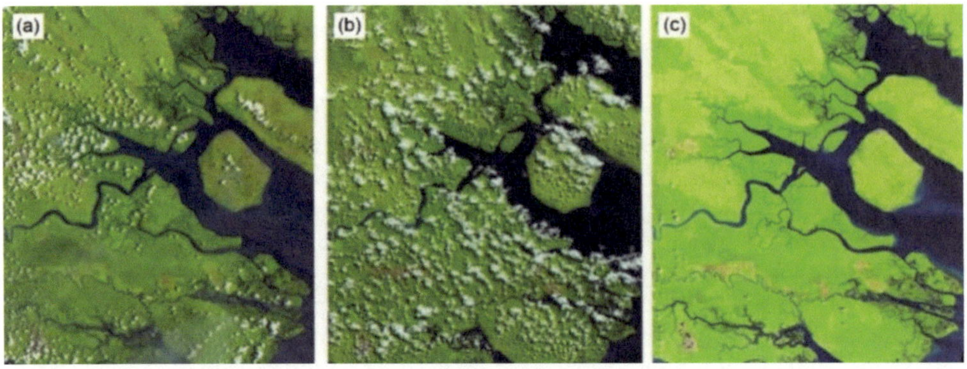

Figure 2.
Cloud detection and removal process. Individual Landsat scene that was captured on 26 January 2017 (a) was merged with that captured on 14 June 2017 (b), where both produced a cloud-free images for the year 2017 (c).

2.4 Images classification

Appropriate enhancement techniques were applied to the images to make the mangroves appear better on the images [17]. In addition to the individual spectral bands of Landsat images, vegetation indices such as Normalized Different Vegetation Index (NDVI), Green Atmospherically Resistant Index (GARI), and Normalized Difference Infrared Index (NDII) were also derived from the images to improve quality of classification. The vegetation indices that were used in this study are summarized in **Table 1**.

Most spectral-based image classifications are performed using traditional methods such as maximum likelihood, linear discriminant analysis, and spectral angle mapper classifiers. These methods are applied to the spectral bands to produce a classified feature in images [18]. Instead of using these approaches, this study attempted a new approach to classify the images. R Package, which is free, open source software with the RandomForest algorithm [19] was used.

RandomForest implements Breiman's RandomForest algorithm, based on Breiman and Cutler's original FORTRAN code for classification and regression [20]. It can also be used for assessing proximities among data points without necessarily a training set. All sampling points that were collected on the ground were connected to the corresponding pixels on the image through this algorithm. Classification was done by searching the most important variables i.e. which spectral bands are used in decision tree approach [21–23]. RandomForest applies four major steps of looking at the importance of variables as follow:

1. Step 1: to determine the significance of the mth variable. In the left out cases for the kth tree, randomly permute all values of the mth variable. Put these new covariate values down the tree and get classifications.

2. Steps 2 and 3: for the nth case in the data, its margin at the end of a run is the proportion of votes for its true class minus the maximum of the proportion of votes for each of the other classes. The 2nd measure of importance of the mth variable is the average lowering of the margin across all cases when the mth variable is randomly permuted as in Step 1. Step 3 then count the margins that was shrank.

3. Step 4: the splitting criterion used in RandomForest is the Gini criterion, a mechanism that can measure the most to least importance of variables used in

Vegetation indices	Formula	Description
NDVI	$NDVI = \frac{NIR - R}{NIR + R}$	Commonly used to delineate vegetation from other features on images and to measure vegetation vigor. It is sensitive to atmospheric effects
GARI	$GARI = \frac{NIR - [G - 1.7 * (B - R)]}{NIR + [G - 1.7 * (B - R)]}$	Normally used for detection of green pigment concentration and differentiate chlorophyll levels. It is more sensitive to chlorophyll concentrations than the atmospheric effects
NDII	$NDII = \frac{NIR - MidIR}{NIR + MidIR}$	It uses near- and mid-infrared bands to detect changes in plant biomass and water stress in wetlands like mangroves

Note: NIR = near infrared, G = green, B = blue, R = red, and MidIR = middle wave infrared channels.

Table 1.
Vegetation indices that were used derived from the images.

decision tree. At every split, one of the mth variables is used to form the split and there is a resulting decrease in the Gini. The sum of all will decrease the forest due to a given variable, normalized by the number of trees.

All images have been classified to distinguish mangroves from the other land uses. The classification results were transformed into vector shapefile for further refinement and editing. The accuracy of the classification results were assessed by using a number of ground truth points. The GIS platform was used to carry out post-classification analysis. Post-classification analysis is usually used for quantifying changes of land uses. Changes of mangroves were identified from the conversions of mangroves to other landuse classes, which are (i) urban, settlement, and industrial areas, (ii) agricultural, (iii) aquaculture activities, and (iv) coastal erosion.

2.5 Estimation of CO_2 emission

Carbon dioxide (CO_2) is defined as natural, colorless and odorless greenhouse gas that is emitted when fossil fuels (i.e. natural gas, oil, coal, etc.) are burnt. In this study, the CO_2 emission is expressed as C loss, assuming that the gas is emitted when deforestation occur. The units of metric tons C was converted to CO_2 by multiplying the ratio of the molecular weight of carbon dioxide to that of carbon (44/12 = 3.67) [24].

The CO_2 resulted from deforestation is one of the important elements in greenhouse gases emissions. Therefore, it is also essential to quantify the contribution of mangrove deforestation towards the CO_2 emission. Net emission as resulted from deforestation of mangroves can be estimated based stock-difference method, which can be expressed as Eq. (1) as follow [24];

$$\Delta C = \frac{(C_{t1} - C_{t2})}{(t_2 - t_1)} \quad (1)$$

where ΔC is changes in carbon stock (Mg C yr^{-1}), C_{t1} and C_{t2} (Mg C) is carbon stock at time t_1 and t_2 (year), respectively. In this case, the C_{t1} and C_{t2} was quantified from the changes analysis that have been carried out earlier this study.

3. Mangroves mapping and monitoring in Malaysia

3.1 Production of seamless image mosaic

F mask algorithm successfully removed almost 100% of cloud covers and their shadows on the images. The algorithm also managed to detect thin, low temperature clouds in the high altitude by thermal sensors onboard the Landsat TM, ETM+ and OLI. The algorithm somehow failed to detect small scattering clouds that occurred in small patches on the images. Nevertheless, the algorithm has facilitated the cloud removal process and make the mangroves mapping and monitoring work at landscape-level practical. **Figure 2** shows a portion of mangroves on two different images that were captured on different dates with clouds. These images were used to produce seamless mosaic of images without cloud covers.

3.2 Identification of mangrove ecosystem and images classification

The study indicated that the suitable spectral bands for species discrimination varied with scale. However, near-infrared (700–1327 nm) bands were consistently important spectrum across all scales and the visible bands (437–700 nm) were more important at pixel and crown scales. By using the RandomForest algorithm, the most important bands in the classification were represented by a mean decrease Gini values. The most important bands in mangroves discrimination, from most to least, are; MidIR, NIR-2, NIR, Green, Blue, Red. Spectral profile of the images also showed that the NIR channels separate the mangroves from the other land covers very well (**Figure 3**). On the other hands, the vegetation indices that were used in this study played similar important role in mangroves classification.

The image classification approach that has been applied in this study was found to be effective only at large coverage of mangroves. The accuracy for all classifications were ranging from 83 to 91%, which were acceptable and reliable for monitoring purpose. Mangroves are normally appear dark on any combination of spectral bands of multispectral image. This is due to the natural ecosystem of mangroves, which is covered by swamps and sometimes inundated by tidal water. The chlorophyll content of the mangrove leaves, which is higher than those of trees and crops, tends to make them appear darker on satellite images [25], as depicted in **Figure 4**. Each mangrove species has a unique configuration of trunks, prop roots and

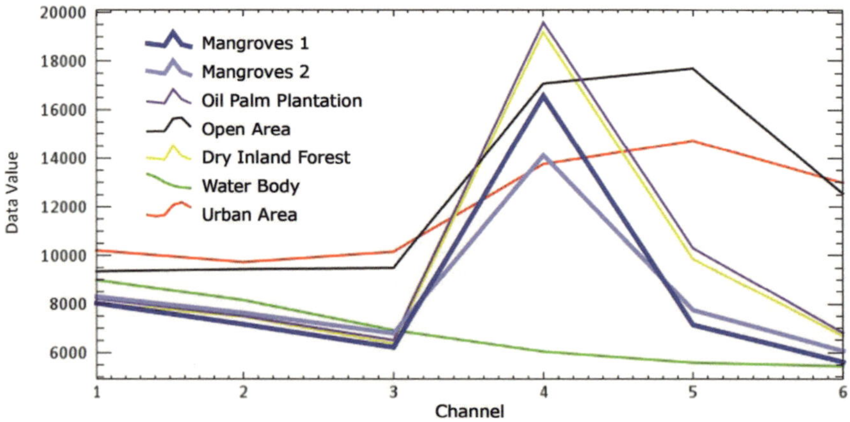

Figure 3.
Spectral profiles of several land covers extracted from the images. Channel 1 through 6 on the y-axis are blue, green, red, NIR, NIR-2 and MidIR, respectively.

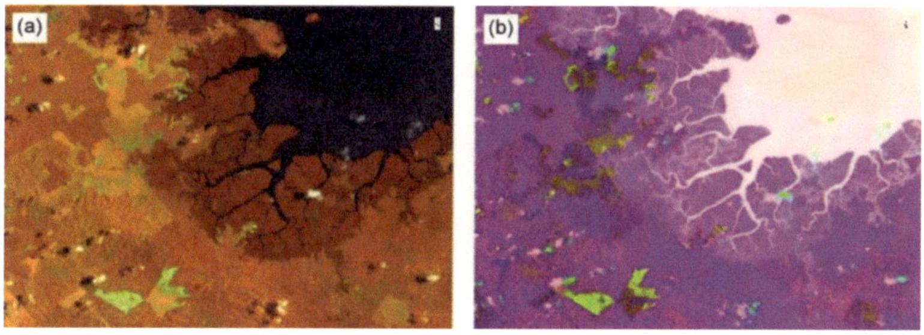

Figure 4.
Images showing (a) combination of bands 5, 6 and 4 of Landsat-8 OLI and (b) combination of vegetation indices, NDVI, GARI and NDII. These images were selected for the classification process.

pneumatophores that works as a different drag force therefore resulting in a different reduction rate of sea waves (**Figure 5**). Not only this, the wet floor of the forest gives special spectral characteristics on satellites images that can be differentiated easily from other features (**Figure 6**).

3.3 Spatial data editing for mangroves mapping

The classification results were further edited to refine the shapes and accuracy. This process was conducted manually on the vector shapefile by visual interpretation on GIS platform. Finally the spatial distribution of the mangroves were mapped properly (**Figure 7**). The mangroves in Malaysia were mostly found in Sabah (60%), followed by Sarawak (22%) and Peninsular Malaysia (18%). **Table 2** summarizes the total extents of mangroves in the respective regions that have been produced from the classification. It is notable that the total extents of mangroves have been decreasing throughout the monitoring period. **Figure 8** shows spatially explicit map of mangroves distribution in Malaysia as of year 2017. Mangroves are found mainly along the west coast of Peninsular Malaysia, west coast of Sarawak and the east coast of Sabah.

3.4 Monitoring of mangroves changes

Table 3 reports the changes of mangroves that occurred over the 27 years of monitoring period. The total loss of mangroves was about 21,274 ha where majority of the mangroves loss were outside the Permanent Forest Reserve or within the

Figure 5.
Roots and successive stands of Rhizophora apiculata in a common mature mangrove forest.

Figure 6.
Mangroves as they appeared on Landsat-8 image. The dark green areas represent the mangrove areas. The image classification process, either automated or manual digitizing, is usually easier for mangrove areas than for other vegetation. The image is displayed using a combination of bands 543 (RGB) over the Kapar area in Klang, Selangor. The central bottom is Klang port complex and the bottom left is Pulau Klang, which is predominantly covered by mangroves.

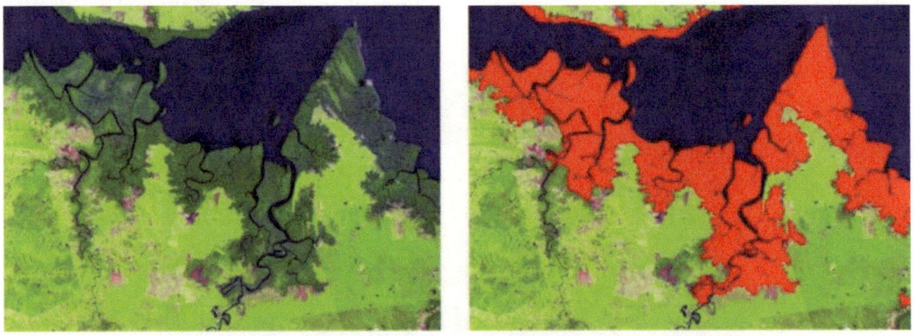

Figure 7.
Mangroves appear dark green on the original image (left) and the classified mangroves, indicated as red polygons (right).

stateland areas. These areas are actually the land bank for the states developments, which are principally included in the State's Structural Planning. Example of mangroves changes detected from the multi-temporal mapping process is shown in

Region	Mangroves 1990 (ha)	Mangroves 2000 (ha)	Mangroves 2017 (ha)
Peninsular Malaysia	116,746	114,353	110,953
Sabah	385,630	382,448	378,195
Sarawak	147,936	145,263	139,890
Total	650,311	642,063	629,038

Table 2.
Extents of mangroves in Malaysia.

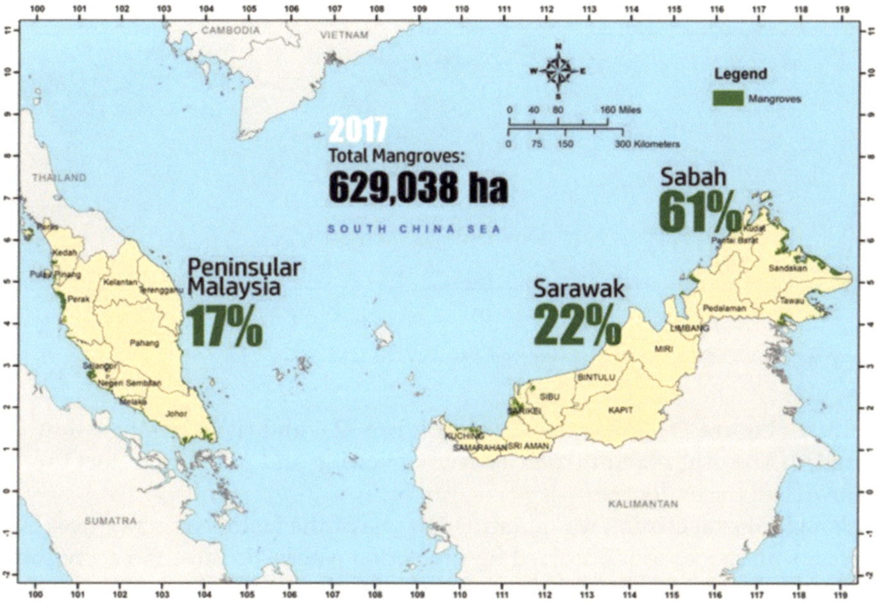

Figure 8.
Distribution of mangroves in Malaysia over the year 2017.

Region	Mangrove loss 1990–2000 (ha)	Mangrove loss 2000–2017 (ha)	Rate of deforestation 1990–2000 (ha yr^{-1}) | (% yr^{-1})	Rate of deforestation 2000–2017 (ha yr^{-1}) | (% yr^{-1})	Average rate of deforestation 1990–2017 (ha yr^{-1}) | (% yr^{-1})
Peninsular Malaysia	2393	3400	239 | 0.20	200 | 0.17	215 | 0.19
Sabah	3182	4253	318 | 0.08	250 | 0.07	275 | 0.07
Sarawak	2673	5373	267 | 0.18	316 | 0.22	298 | 0.21
Total	8227	13,190	823 | 0.13	776 | 0.12	793 | 0.13

Table 3.
Mangroves deforestation in Malaysia between years 1990 and 2017.

Figure 9. From this information, it can be concluded that the annual decrease rate of mangroves was about 788 ha per year or about 0.13% per annum since year 1990. Major factors that contributed to these changes have been identified as: (i) direct conversion to other land uses (**Figure 10**), predominantly for commercial-scale

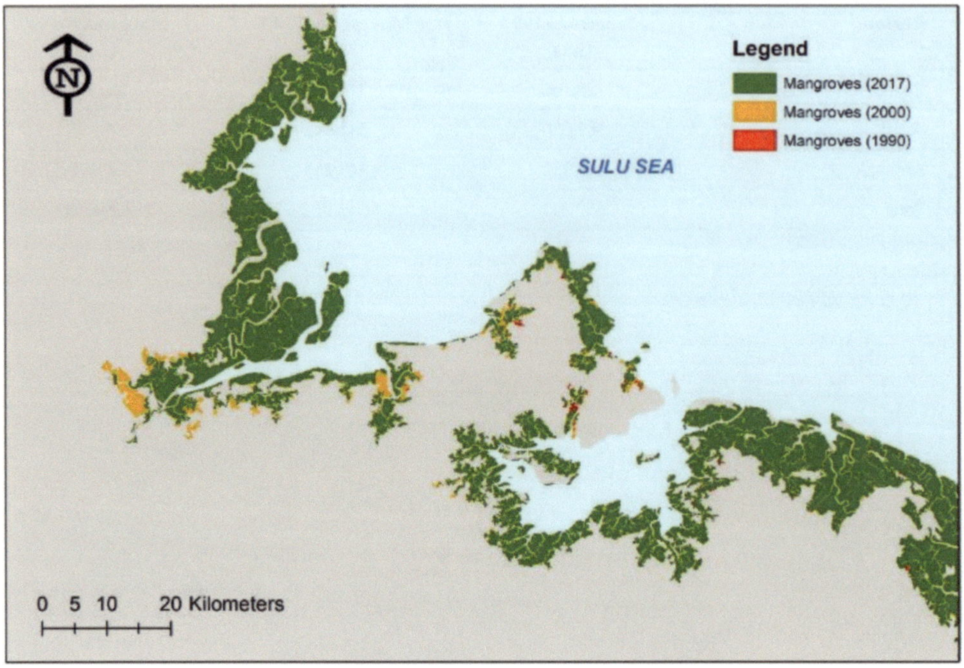

Figure 9.
Changes of mangroves that occurred between 1990 and 2017 overlaid on GIS platform.

agriculture (**Figure 11**) and aquaculture (**Figure 12**), and (ii) coastal erosion (**Figure 13**). The other factors such as overharvesting and pollution affect the mangroves to a lesser degree.

Although coastal erosion was identified as one of the factors of mangroves loss, there were some accretions occurred in some other places. Erosion and accretion is a dynamic process and takes place along the coastlines and major estuaries, where suspended sediments are likely to settle. These phenomena also lead to species succession when the existing plant species die due to unsuitable soil and new species emerge. Besides, mangrove roots can act as wave breaker and promote flocculation and sedimentation, eventually forming mudflats that allow positive accretion (**Figure 14**). Coastal erosion occurs when the waves hit perpendicular to the coastlines and when the rapid flow of sea currents wash away the sand or soil particles. The frequency and height of waves hitting the coastlines contribute to the harshness of coastal erosion. Thus, the presence of mangroves can reduce the coastal erosion significantly. This condition is obvious particularly in the areas facing the sea [26, 27].

Figure 10.
Land developments on mangroves. Reddish color represents newly opened areas for development purposes that were cleared from the original mangroves areas (dark green color).

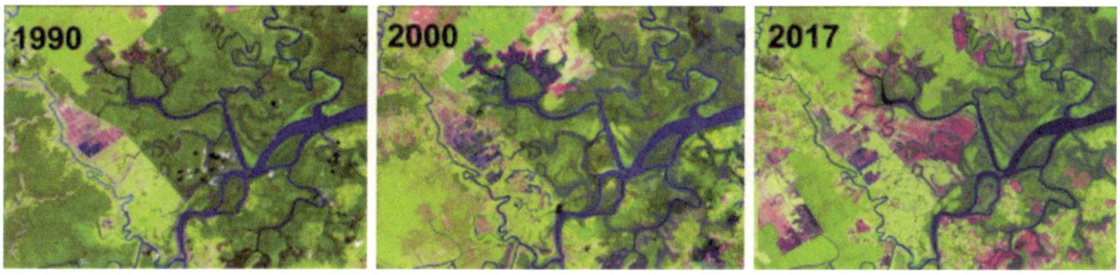

Figure 11.
Expansion of oil palm plantation on mangroves. Reddish color represents newly opened plantations from the original mangroves areas (dark green color). The bright green represents existing plantations.

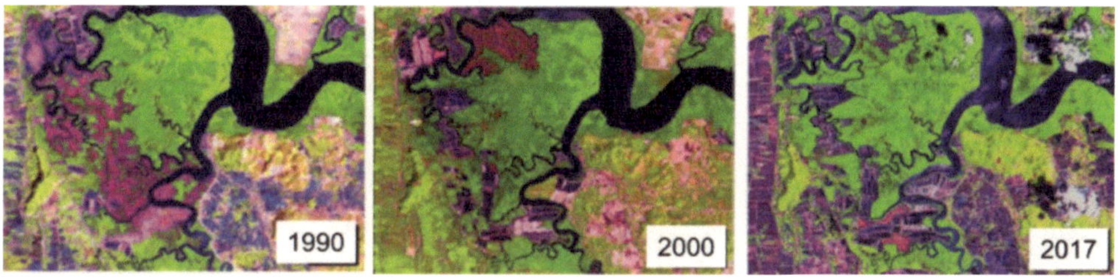

Figure 12.
Expansion of aquaculture industries on mangroves. Dark blue patches represents newly opened aquaculture ponds from the original mangroves areas (dark green color).

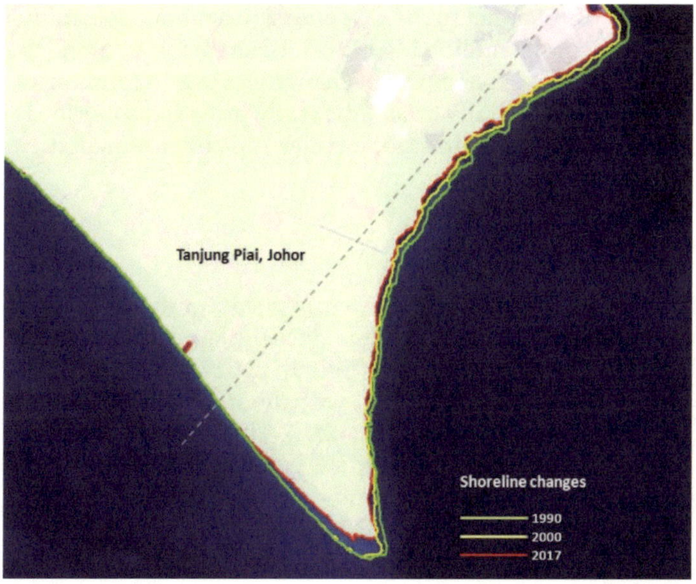

Figure 13.
Shoreline changes that resulted from coastal erosion along the coast of south Pontian, Johor. The study indicated that 14.2 km stretches have been facing serious coastal erosion within the last two decades with the rate of erosion ranging from 3.2 to 12.5 m per year.

3.5 The estimated carbon emission

A study has indicated that the average C stock (aboveground and belowground) in mangroves in Malaysia is about 181 Mg C ha^{-1} [28]. The extents of mangroves loss for each epoch were multiplied by this average carbon stocks. The study demonstrated that the total loss of carbon due to the loss of mangroves was about 2.6 million

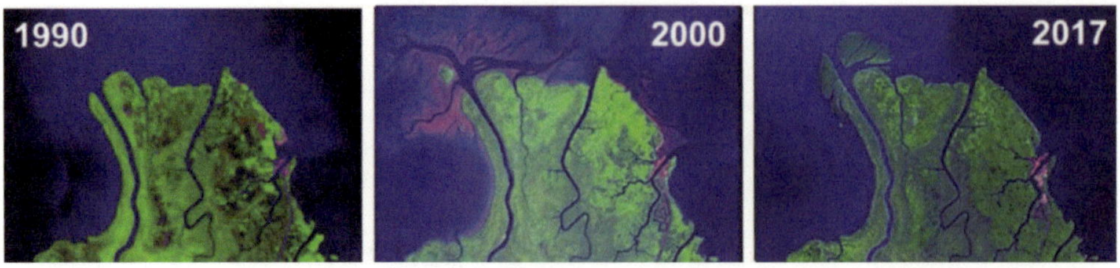

Figure 14.
Positive accretion of mangroves at estuaries. The new formations at the river mouths were colonized by mangroves trees forming a naturally generated forest.

Region	Mangrove loss (ha)	Carbon loss (Mg C)	CO_2 emission (Mg CO_2)	Rate of CO_2 emission (Mg CO_2 yr^{-1})
Peninsular Malaysia	5793	1,048,567	3,848,242	142,527
Sabah	7435	1,345,672	4,938,617	182,912
Sarawak	8046	1,456,288	5,344,578	197,947
Total	21,417	3,876,409	14,226,422	526,905

Table 4.
CO_2 emission resulted from mangroves loss between years 1990 and 2017.

Mg C. Subsequently, this has led to the CO_2 emission at about 14.2 million Mg CO_2, with an average of about 0.5 million Mg CO_2 emission per year, along the monitoring period. **Table 4** summarizes the impact of mangroves loss in terms of CO_2 emission. Although the figures are generally crude, the study provided some ideas for further studies, especially which related to carbon cycles and climate change.

4. Conclusion

This study has successfully assessed the current state of mangroves and determined the rate of mangroves loss in Malaysia since the last decade. Total mangroves in Malaysia has decreased from 650,311 ha in 1990 to 629,038 ha in 2017. Total deforestation was accounted at 21,274 ha or 3.3% with the annual rate of deforestation of 788 ha yr^{-1} or 0.13% yr^{-1}, between 1990 and 2017. The study also quantified the C stock changes and estimated CO_2 emission due to the loss of mangroves in Malaysia. Total emission caused by the mangroves deforestation was accounted at about 14 million Mg CO_2 with annual emission rate of around 0.5 million Mg CO_2 yr^{-1}.

The study found that the Landsat-based mapping and monitoring of mangroves was very practical. It provides a reliable information on mangroves distribution, both qualitatively and quantitatively. Landsat missions provide a very useful RS tool for monitoring changes of mangroves over time. The study suggests that appropriate actions should be taken by the Government of Malaysia to protect the mangroves and keep their ecosystem intact forever. The most effective way to conserve the mangroves is to gazette the remaining stateland forest as Permanent Reserved Forests (PRFs). These PRFs should then be maintained as amenity for current and future generations, while contributing to the mitigation of climate change impacts at the local level. Any development in PRFs should be prohibited or implemented with caution.

Overall, there is great potential in the application of Landsat-based data with appropriate GIS technique for mapping and monitoring of mangroves in Malaysia.

Although there are cloud covers problems on some of the images, this has not hindered the assessment of mangroves at landscape and regional levels. The accuracy and precision also vary depending on the objective of the application. However, the ability to detect major changes in the ecosystem that can cause profound and irreversible damage far outweighs a perfectly or highly accurate and precise RS based method at this point.

Currently, Malaysia has reserved about 85% (~535,000 ha) out of the total areas of mangroves as Permanent Forest Reserve and State/National Parks. The remaining 15% is under the state-lands and alienated lands. By far, the most effective way to preserve these mangroves is through gazzeting into permanent forest reserves.

Acknowledgements

This work has been carried out under the Research and Development Committee on Mangroves (JTRD) led by FRIM. Special thanks for the Forestry Department Peninsular Malaysia (JPSM), Sabah Forestry Department (SFD), and Forest Department Sarawak (FDS) for the supports on the ground data collection activities.

Conflict of interest

The authors declare no 'conflict of interest' for this chapter.

Author details

Hamdan Omar*, Muhamad Afizzul Misman and Samsudin Musa
Forest Research Institute Malaysia, Kepong, Selangor, Malaysia

*Address all correspondence to: hamdanomar@frim.gov.my

References

[1] Daniel CD, Kauffman JB, Murdiyarso D, Kurnianto S, Stidham M, Kanninen M. Mangroves among the most carbon-rich forests in the tropics. Nature Geoscience Letters. 2011;**4**:293-297

[2] International Union for Conservation of Nature (IUCN). IUCN Red List of Threatened Species™. Version 2011.1. 2011. Available at: www.iucnredlist.org

[3] Giri C, Long J, Abbas S, Murali RM, Qamer FM. Distribution and dynamics of mangrove forests of South Asia. Journal of Environmental Management. 2015;**148**:101-111

[4] Azahar M, Nik M, Shah NM. A Working Plan for the Matang Mangrove Forest Reserve, Perak: The Third 10-Year Period (2000-2009) of the Second Rotation. Ipoh: State Forestry Department of Perak; 2003

[5] Roslan A, Shah NM. A Working Plan for the Matang Mangrove Forest Reserve, Perak: The First 10-Year Period (2010-2019) of the Third Rotation. Ipoh: State Forestry Department of Perak; 2014

[6] Food and Agriculture Organization (FAO). Asia-Pacific forestry sector outlook study. Country report—Malaysia. In: Working Paper Series (FAO):7/FAO Bangkok (Thailand). Regional Office for Asia and the Pacific; 1997

[7] Aizpuru M, Achard F, Blasco F. Global assessment of cover change of the mangrove forests using satellite imagery at medium to high resolution. EEC Research Project No. 15017-1999-05 FIED ISP FR. Ispra, Italy: Joint Research Centre; 2000

[8] Food and Agriculture Organization (FAO). The World's Mangrove 1980-2005. FAO Forestry Paper 153. Rome: FAO; 2007

[9] Spalding M, Kainuma M, Collins L. World Atlas of Mangroves. UK: Earthscan; 2010

[10] Giri C. Observation and monitoring of mangrove forests using remote sensing: Opportunities and challenges. Remote Sensing. 2016;**8**(783):1-8

[11] Cárdenasa N, Karen Y, Joycea E, Maiera SW. Monitoring mangrove forests: Are we taking full advantage of technology? International Journal of Applied Earth Observation and Geoinformation. 2017;**63**:1-14

[12] Sulong I, Mohd-Lokman H, Mohd-Tarmizi K, Ismail A. Mangrove mapping using Landsat imagery and aerial photographs: Kemaman District, Terengganu, Malaysia. Environmental Development and Sustainability. 2002;**4**:135-152

[13] Khali Aziz H, Hamdan O, Shamsudin I, Ismail H. Digital change detection of mangrove forest in Selangor using remote sensing and geographic information system (GIS). The Malaysian Forester. 2009;**72**(1):61-69

[14] Hamdan O, Khali Aziz H, Shamsudin I, Raja Barizan RS. Status of Mangroves in Peninsular Malaysia. Kepong, Selangor: Forest Research Institute Malaysia; 2012. p. 134

[15] Zhu Z, Woodcock CE. Object-based cloud and cloud shadow detection in Landsat imagery. Remote Sensing of Environment. 2012;**118**:83-94

[16] Zhu Z, Wang S, Woodcock CE. Improvement and expansion of the Fmask algorithm: Cloud, cloud shadow, and snow detection for Landsats 4-7, 8, and Sentinel 2 images. Remote Sensing of Environment. 2015;**159**:269-277

[17] Hesham AM, Smith SE. A new approach for estimating mangrove

canopy cover using Landsat 8 imagery. Computer and Electronics in Agriculture. 2017;**135**:183-194

[18] Matthew LC, Roberts DA, Clark DB. Hyperspectral discrimination of tropical rain Forest tree species at leaf to crown scales. Remote Sensing of Environment. 2005;**78**:24-47

[19] Liaw A, Wiener M. Classification and regression by RandomForest. R News. 2002;**2**:18-22

[20] Breiman L. Manual on Setting Up, Using, and Understanding RandomForests V3.1. 2002. Available from: http://oz.berkeley.edu/users/breiman/Using_random_forests_V3.1.pdf [Accessed: 21st Nov 2010]

[21] Lawrence RL, Wood SD, Sheley RL. Mapping invasive plants using hyperspectral imagery and Breiman cutler classifications (RandomForest). Remote Sensing of Environment. 2006;**100**(3):356-362

[22] Adam E, Mutanga O, Rugege D, Ismail R. Discriminating the papyrus vegetation (*Cyperus papyrus* L.) and its co-existent species using RandomForest and hyperspectral data resampled to HYMAP. International Journal of Remote Sensing. 2012;**33**(2):552-569

[23] Mansour K, Mutanga O, Everson T, Adam E. Discriminating indicator grass species for rangeland degradation assessment using hyperspectral data resampled to AISA Eagle resolution. ISPRS Journal of Photogrammetry and Remote Sensing. 2012;**70**:56-65

[24] Intergovernmental Panel on Climate Change (IPCC). In: Eggleston HS, Buendia L, Miwa K, et al., editors. IPCC Guidelines for National Greenhouse Gas Inventories; Prepared by the National Greenhouse Gas Inventories Programme. Hayama, Japan: Japan Institute for Global Environmental Strategies; 2006

[25] Hamdan O, Norsheilla MJC, Ismail P, Samsudin M, Norhaidi Y. Assessing the rate of deforestation and changes of carbon stock on mangroves in Pahang, Malaysia. The Malaysian Forester. 2016;**79**:174-179

[26] Khairuddin B, Yulianda F, Kusmana C, Yonvitner. Degradation mangrove by using Landsat 5 TM and Landsat 8 OLI image in Mempawah regency, West Kalimantan Province year 1989-2014. Procedia Environmental Sciences. 2016;**33**:460-464

[27] Kanniah K, Sheikhi A, Cracknell A, Goh H, Tan K, Ho C, et al. Satellite images for monitoring mangrove cover changes in a fast growing economic region in southern peninsular Malaysia. Remote Sensing. 2015;7:14360-14385

[28] Ibharim NA, Mustapha MA, Lihan T, Mazlan AG. Mapping mangrove changes in the Matang mangrove Forest using multi temporal satellite imageries. Ocean & Coastal Management. 2015;**114**:64-76

Section 4

Big Data and Augmented Reality

Chapter 7

GIS and Big Data Visualization

Junghoon Ki

Abstract

Geographic information system (GIS) has expanded its area of applications and services into various fields, from geo-positioning service to three dimensional demonstration and virtual reality. Big data analysis and its visualization tools boosters the capacity of GIS, especially in graphics and visual demonstration. In this chapter, I describe major traits of big data and its spatial analysis with visualization. And then I will find a linkage between big data and GIS. There are several GIS-based software and geo-web that deal with big data or similar scaled databases, such as ArcGIS, Google Earth, Google Map, Tableau, and InstantAtlas. For these software and websites are developed based on geography or location, they still have some limits in visualizing big data or persuading people with maps or graphics. I will search a way out of this limitation of GIS-based tools and show an alternative way to visualize big data and demonstrate thematic maps. This chapter will be a useful guide to lead GIS people into a new horizon of big data visualization.

Keywords: GIS, big data, visualization, infographic, Tableau, Chernoff face

1. Introduction

For decades, geographic information system (GIS) has expanded its area of applications and services into various fields, from geo-positioning service to three dimensional demonstration and virtual reality. It is a tremendous progress of GIS since its burgeoning as a combination of map and database. Today, everyone in the world is living, working, and resting under the umbrella of GIS applications and services in the form of navigation system, the Google Earth, GPS, and even Pokémon GO.

Stronger and more fundamental changes are asked in GIS development when big data emerged in the early 2010 [1, 2] (see **Figure 1**). Characterized with a large volume, a vast variety, and a fast velocity, big data has been releasing the explosive datasets in social media and other complex platforms. Is big data just good news to the GIS community? According to Sanderson [3], there are still some hurdles that preventing GIS and big data from joining together. They are related with big data's unstructured data structure, real time data production, accuracy, and scale. Not only these obvious limitations of big data, it also neglects locations of datasets frequently. Big data deals mainly with information, not necessarily geography.

What is a pivotal role of big data in GIS development? It is necessary to take a look at a process of big data production to find out a connection between two fields. The big data production process consists of data collection, storage, computing & batching, analysis, and visualization & demonstration. Among the process, visualization and demonstration could provide an effective and efficient way with GIS people in terms of new interpretation and creative advertisement.

Figure 1.
Google Trends Search for big data since 2004 to present.

Big data's several or more visualization tools with their software are creating a lot of wonderful GIS masterpieces recently. Thereafter, I examine those tools and find some implications from them. Can big data visualization overcomes the limitations of GIS and opens a new horizon? This chapter would provide answers to this question.

The followings constitute sub-sections of the chapter.

- What is big data
- Big data and geographic information system (GIS)
- Big data as an alternative visualization tool for GIS
- Can big data visualization overcome GIS limitations?
- Conclusion

2. What is big data?

2.1 Big data's characteristics and components

Big data can be defined as datasets which have various data styles, fast processing speed, and are hard to be managed and analyzed with existing data systems. These characteristics of big data are summarized with '3V', which denotes volume, variety, and velocity [4].

First, big data deals with large volume datasets, usually more than terabyte size that usually comes from Global Positioning System (GPS), social media, and other sensors. A terabyte is a unit of information equal to one million * million (10^{12}) bytes, or 1024 gigabyte. The brand 'big data' itself implies a size of datasets is very huge compared to past datasets.

Second, big data deals with a variety of datasets such as sound, picture, video stream, map and even social media text message. Big data targets not only structured datasets but also unstructured ones that were usually out of interest to data workers. Its range is beyond our imagination and different kinds of datasets are integrated to generate new types of database. Big data systems use a computer clouding and other platform such as Hadoop for data combination and integration (see **Figure 2**).

Big data's third characteristic is velocity because it's very fast in generating, spreading, and applying in the real world. Big data's speed in generation, spread, and application can be accelerated with social media or social network services such as Facebook or Twitter [6]. When people post photos in Facebook, those are recorded as datasets, which offer the useful real-time evidence of locations, preference, and other personal information (see **Figure 3**). This information will be used for marketing and sales by private business or policy measures by government sector.

Although a narrow definition of big data emphasizes data source, collection, storage and other technical issues, its wider definition embraces analysis and demonstration aspects. In summary, big data is defined as very large-sized, various-formatted datasets and analytic methods based on engineering technology and social network services, including statistical fusion and new visualization.

Major components of big data are resource, technology, and human capital [4]. Resource here indicates data acquisition and quality management. Big data technology denotes its platform that refers to data storage, management, processing, analysis, and visualization. Human capital in big data is called data scientists who have an ability of mathematics, engineering, economics, statistics, and psychology. They are also asked to have a capacity of communicating with other people, making a creative storytelling, and visualizing their big data contents effectively.

2.2 Big data's data process and analysis techniques

In **Figure 4**, I briefly show big data process with its elements in which the process has data source, collection, storage, processing, with analysis and visualization. Each step of process has a considerably different elements from the past database systems that generally dealt with structured datasets.

First, big data's data sources come from institutions' or organizations' internal database, or external database such as Twitter or Facebook, or pictures and video streams. Generally, urban and geographic researches and projects use a large scale spatial database [7], which can be called big data.

Second, in the collection process, big data utilizes a crawling method with search engine to get Internet data. It also uses Internet of Things (IoT) based sensors to collect data. This step makes a huge difference to big data from the past data collection traditions.

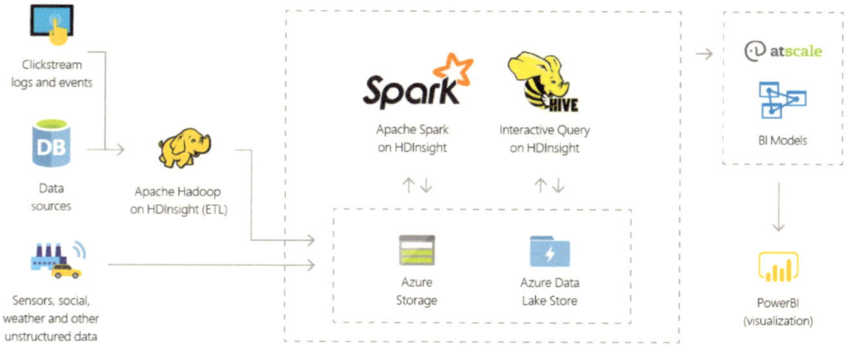

Figure 2.
Data warehousing with Apache Hadoop [5].

Figure 3.
Twitter image that contains some personal information.

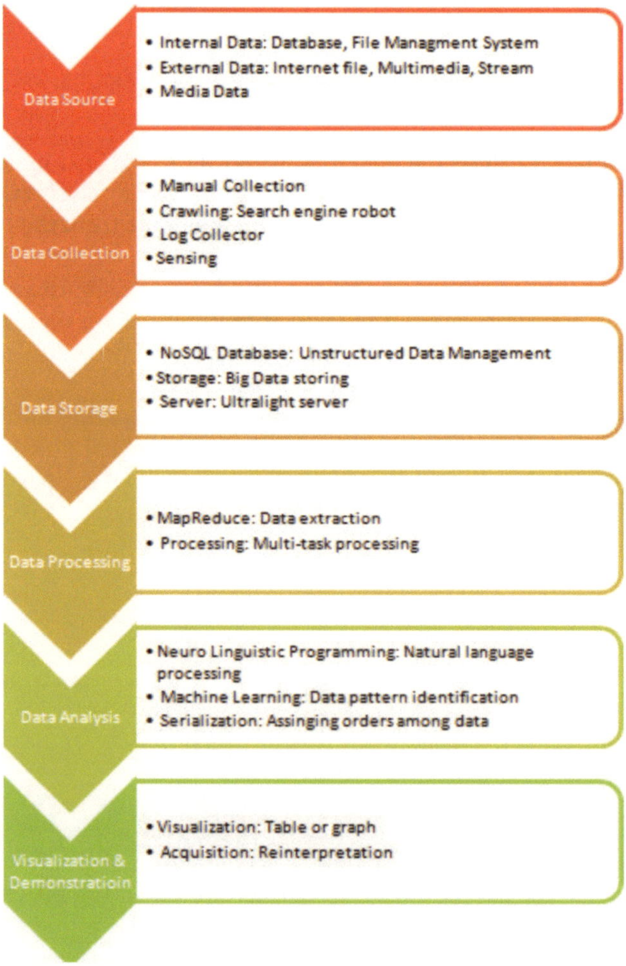

Figure 4.
Big data processing and elements [4, 9].

Third, data storage is a step that engineering technologies are concentrated. Big data managers have to control unstructured data with Not Only SQL (NoSQL), extract data with MapReduce, and execute a distributed parallel processing with Hadoop.

In big data analysis, researchers use neurolinguistic programming for natural language processing, machine learning for data pattern identification, and serialization for assigning orders among data. Researchers pay attention to R programming to conduct big data analysis because it is an efficient statistical tool compared to other packages. Many statistical packages begin to equip themselves with big data analysis modules recently.

Big data visualization and demonstration is a process that analyzed datasets are expressed with graph or table format. Merits of big data visualization in comparison with traditional data visualization is that the former uses word/text/tag clouds, network diagrams, parallel coordinates, tree mapping, cone trees, and semantic networks [8] more often than the latter because its data source format and their needs. R, Tableau, Python language are getting a new attention as effective visualization tool for big data demonstration.

In the next section, I find out a relationship between big data and GIS in terms of these six steps of big data processing.

3. Big data and geographic information system

Big data and GIS are able to share several aspects together because they are similar in elements of data processing. In **Figure 5**, I show GIS data processing with its elements. There are popular open source or commercialized software and web-based online GIS systems, which play an important role in processing and analyzing GIS data.

First, GIS uses data that contains a location or space, therefore it is displayed in a map or picture form. Recently, aerial or satellite data becomes more and more important as new technologies are introduced. As a location based data, GIS data is usually large-sized as is big data.

Second, GIS collects field data such as street information, Closed Circuit TV (CCTV), or other location-based datasets. If the datasets do not provide location information, GIS technicians should perform a geo-coding process to convert into GIS datasets. People's participation is also an important way to get GIS data; so the participatory GIS system becomes a significant field of GIS. Crawling with search engine robot is also useful tool for obtaining data in GIS.

Third, GIS has web server, geospatial data server, or cloud server for its data storage. These servers can be overlapped one another sometimes, but they have their own territories that cannot be shared. In **Figure 6**, I introduce a basic principle of geo-database for single-user and multi-users with the ESRI's official website information. Geo-database system is crucial to manage complicated structured GIS datasets and their attributes.

Fourth, GIS desktop and online software plays a pivotal role in the rest of process including data processing (building), analysis, and visualization. In the GIS data processing (building), efficient systems included are ArcGIS Online, Google Maps JavaScript API, Here Maps JavaScript API, Microsoft Bing Geocode Dataflow API, and US Census Geocoder. They are helpful for building up geo-coding and mapping coordinates in the database.

Fifth, GIS data analysis contains several functions as **Table 1** briefly shows with ArcGIS analysis toolbox summary. Similar analyses are conducted with other software such as ArcGIS, QGIS, GRASS GIS, GeoDa, CartoDB, Mapbox, and the other desktop or online GIS systems.

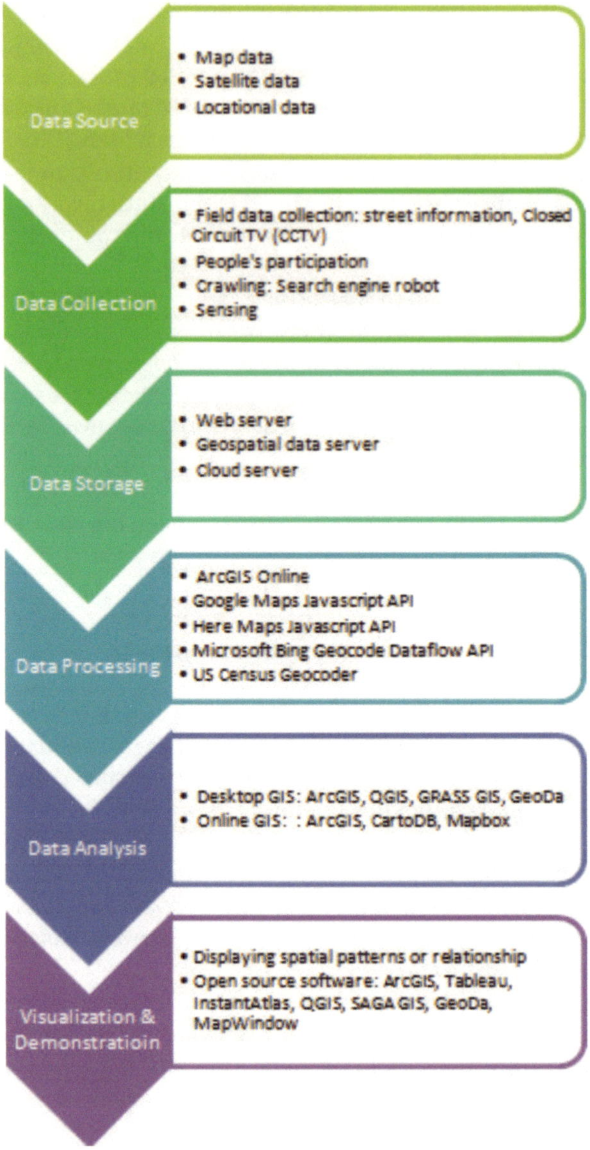

Figure 5.
GIS data processing and its elements.

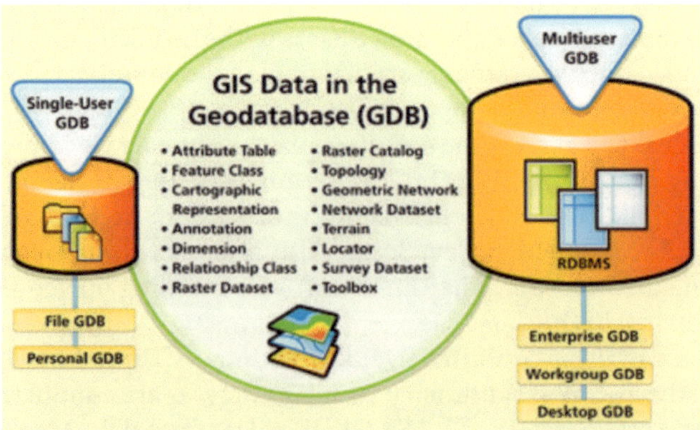

Figure 6.
GIS data in the geo-database (GDB) [10].

Toolsets	Description
Extract	GIS datasets often contain more data than you need. The Extract tools let you select features and attributes in a feature class or table based on a query (SQL expression) or spatial and attribute extraction. The output features and attributes are stored in a feature class or table.
Overlay	The Overlay toolset contains tools to overlay multiple feature classes to combine, erase, modify, or update spatial features, resulting in a new feature class. New information is created when overlaying one set of features with another. There are six types of overlay operations; all involve joining two existing sets of features into a single set of features to identify spatial relationships between the input features.
Pairwise Overlay	The Pairwise Overlay toolset provides an alternative to some of the tools in the Overlay toolset.
Proximity	The Proximity toolset contains tools that are used to determine the proximity of features within one or more feature classes or between two feature classes. These tools can identify features that are closest to one another or calculate the distances between or around them.
Statistics	The Statistics toolset contains tools that perform standard statistical analysis (such as mean, minimum, maximum, and standard deviation) on attribute data as well as tools that calculate area, length, and count statistics for overlapping and neighboring features. The toolset also includes the Enrich tool that adds demographic facts like population or landscape facts like percent forested to your data.

Table 1.
GIS data analysis toolbox example with ArcGIS [11].

Sixth, GIS data visualization intends to display spatial patterns or relationship between or among locations. Popular open source software included here are ArcGIS, Tableau, InstantAtlas, QGIS, SAGA GIS, GeoDa, and MapWindow. These tools are

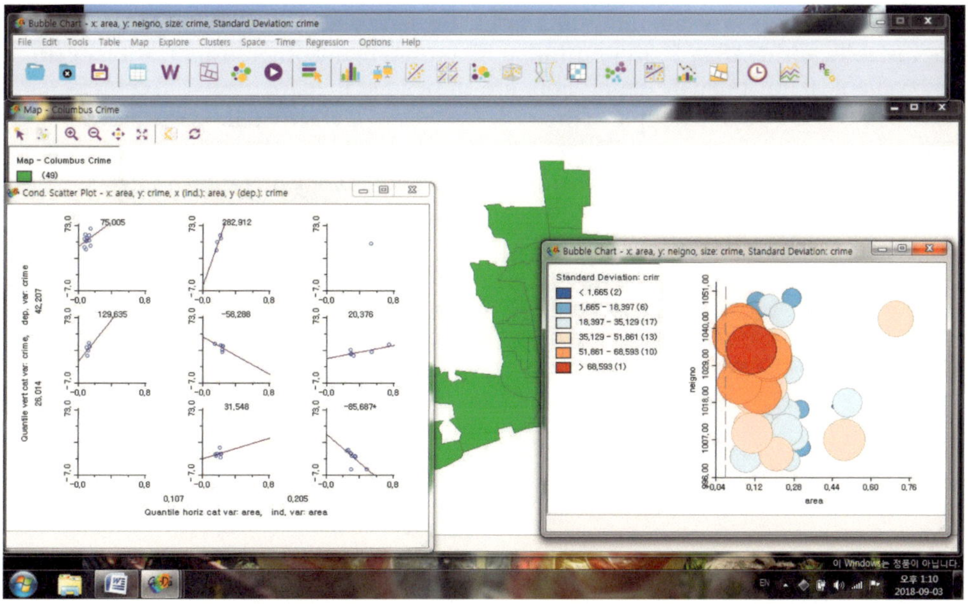

Figure 7.
GeoDa desktop software's visualization windows with crime data.

actively adapted to big data based software or systems to build up location oriented systems as well as more persuasive graphic works. **Figure 7** shows visualization windows with GeoDa desktop software.

4. Big data as an alternative visualization tool for GIS

Can big data be an alternative tool for visualizing GIS and mapping works? Does big data plus location data equal to GIS data? Does big data visualization have any hidden card that surpasses GIS visualization and mapping? I will find answers to these questions in writing this section. Big data's potential for an alternative visualization tool for GIS is to be drawn from several examples in big data technology.

In the visualization and demonstration technology, big data and GIS share together in some aspects. However, there is each field's original aspect that cannot be shared or come together (see **Figure 8**). In **Figure 8**, there are three areas defined: (A) as a GIS visualization's exclusive area, (C) as a big data visualization's exclusive area, and (B) as an overlapping area between two technologies.

GIS visualization's exclusive area (A) indicates that visualization takes places based on location or map with geographic coordinates. Meanwhile, big data visualization's exclusive area (C) means a visualization demonstration without a location or map, which denotes no spatial context are provided. Many big data visualization outcomes do not have any geographic traits or variables and belong to this exclusive area.

Figure 9 is an exemplary map of the area (A), while **Figure 10** is an instance of the area (C). **Figure 9** shows US cities by their elevation in which larger bubble implies higher the city location. I can create this figure using US city and state shape (.shp) files with ArcView GIS software. **Figure 10** shows a gender and ethnicity in tech companies with online Tableau public. In this visualization, there is no evidence of location or mapping technology. This is a pure big data visualization area that is not related with a spatial context or geographic coordinates.

What is the overlapping area (B) that both GIS and big data work together or cooperate? In the (B) area, locations or geographic coordinates are important factor, and big data visualization technologies are also playing a crucial role in demonstration. In **Figure 11**, I provide an example of area (B) with the Chernoff face and US map, in which the Chernoff face denotes multivariate big data visualization using human face-like variables with SAS or R programming. There are many other visualization examples available if any big data expressions are embedded in maps or spatial context. **Figure 11** is also a good example of area (B) because it is clearly telling the location although it does not use a map. **Figure 12** shows how much population is moving from a continent to another with big data visualization technology of Tableau software.

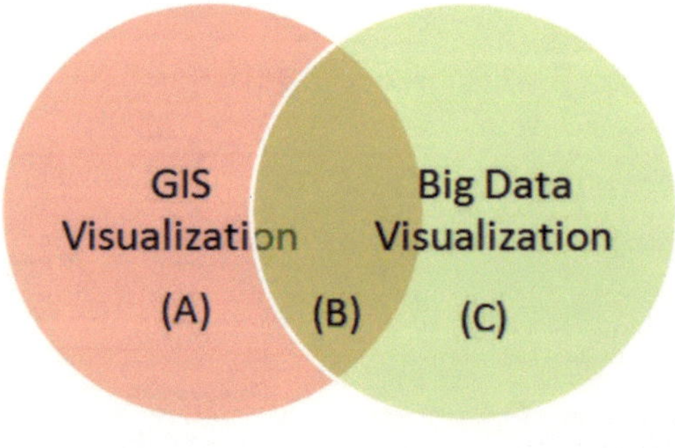

Figure 8.
GIS visualization and big data visualization Venn diagram.

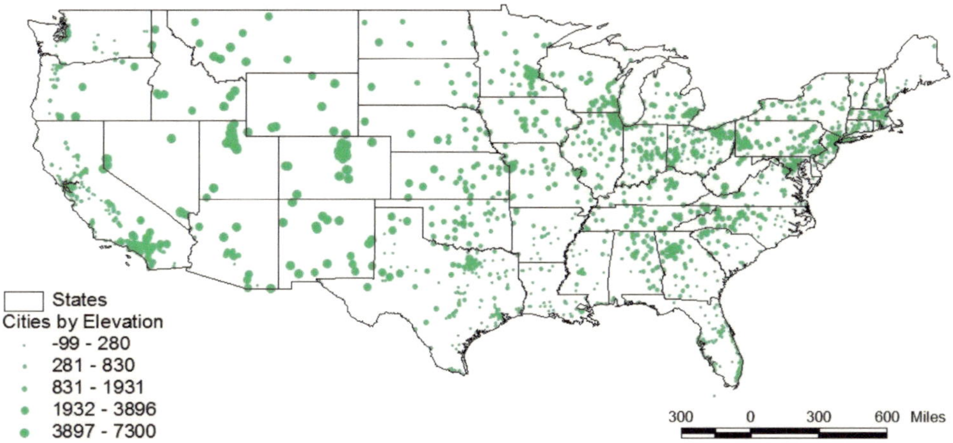

Figure 9.
GIS visualization example: US cities by their elevation with ArcView GIS.

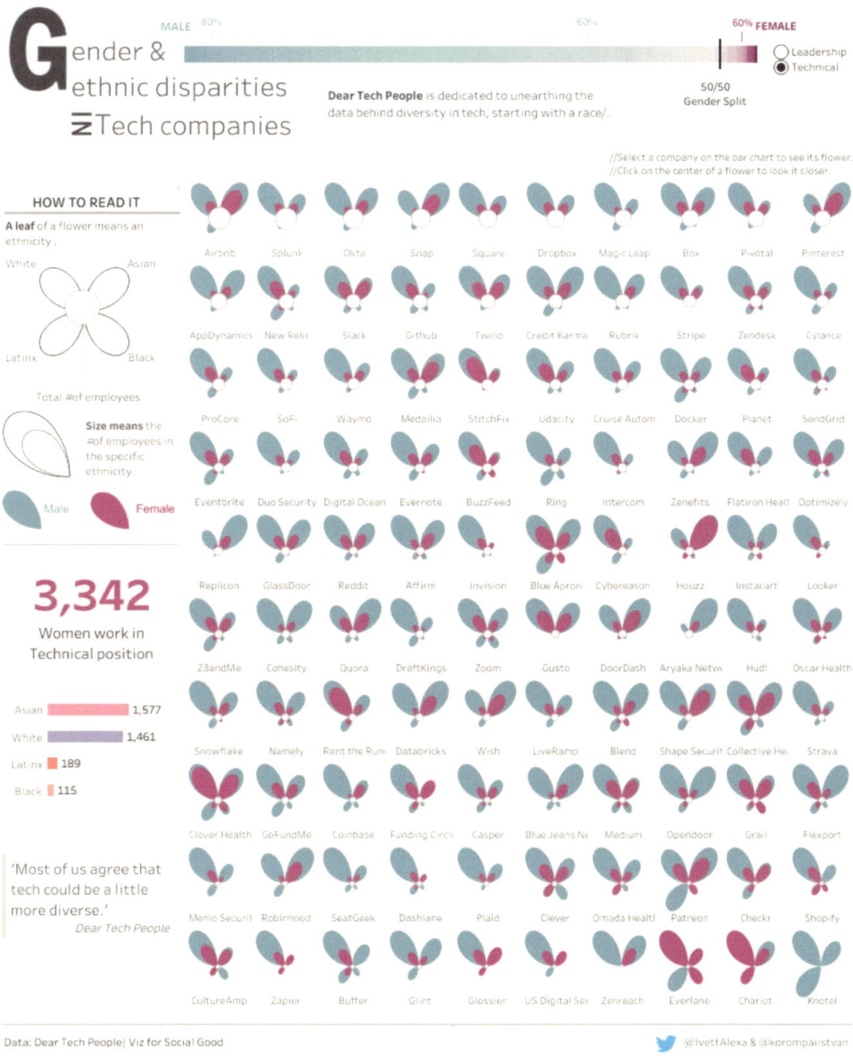

Figure 10.
Big data visualization example: gender and ethnicity in tech companies with online Tableau public [12].

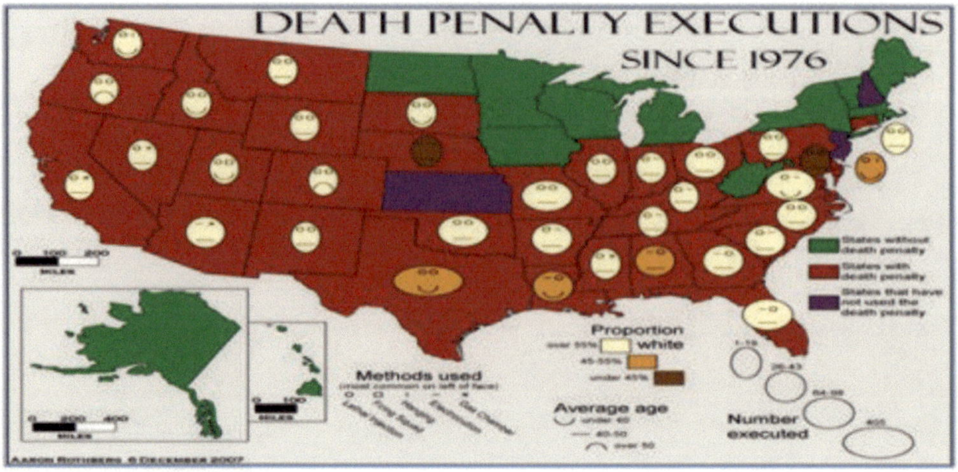

Figure 11.
US states' death penalty executions since 1976 [13].

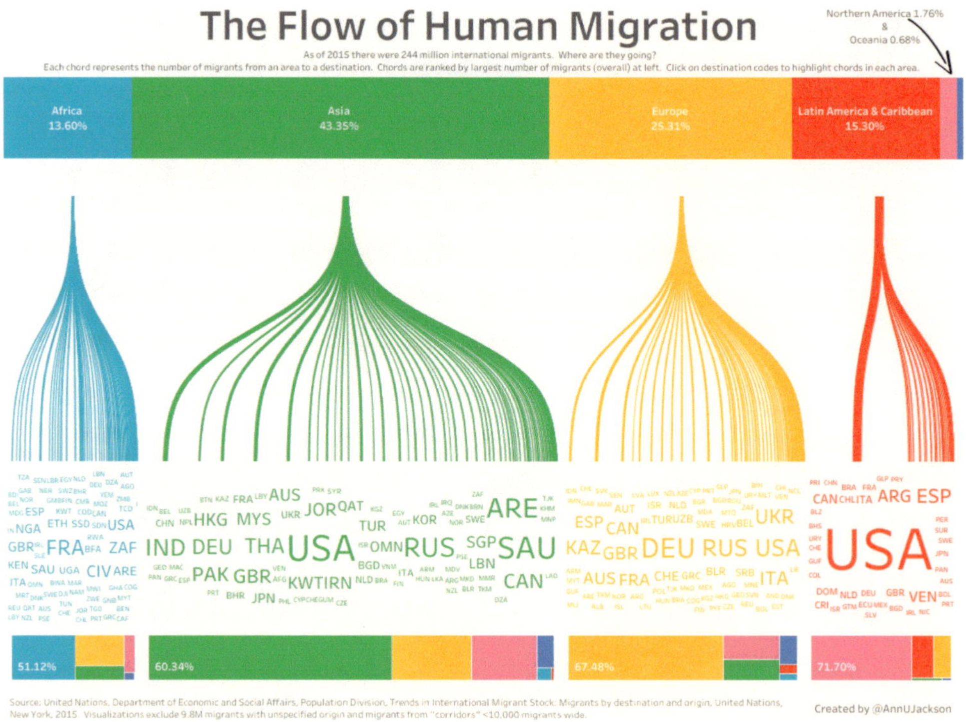

Figure 12.
The flow of human migration with online Tableau public [12].

Does big data visualization overcome GIS and its limitation? About this issue, I describe some insights in the following section.

5. Can big data visualization overcome GIS limitations?

GIS visualization has a limitation since it is basically rooted at the spatial context and geographic maps. GIS visualization's first priority tends more to be geographic than to be informational or graphical. Location matters at GIS visualization as it did at mapping and geography.

Big data visualization opens a new horizon in GIS visualization because it does not just strengthen the spatial context, but also it gives new meanings and insights to GIS maps and demonstration. As is compared in **Figures 8** and **10**, dots in GIS visualization turn into human faces in big data visualization. **Figure 11** implies that locations can be read without a map. More big data visualization skills and their outcomes will be brought out with more abundant insights and implications to GIS visualization.

However, there are some risks of big data visualization in applying to GIS visualization because their fundamental approaches are different in some ways.

First, big data's engineering technologies tend to be ignorant to geographic perspectives. Big data engineers and visual technicians are not necessarily geographers, spatial experts, or even urban planners. Big data visualization workers if loaded with GIS related jobs should be aware of basic spatial principles and mapping process.

Second, GIS experts who is creating big data related visualization should be ready to adapt themselves to engineering guidelines that ask them set their spatial norms aside to set up new GIS-based big data visualization works. When GIS professionals get a step back, they will experience a power of big data visualization technology.

Third, GIS and big data visualization works should be multidisciplinary projects or research, in which all possible fields of study are involved in the final production. Social scientists, data engineers, medical & health experts, graphic designers, and other research fields' professionals can join to generate meaningful GIS visualization performance [14].

Big data visualization can be a good measure if people involved are deliberately designed, called, instructed, and allocated.

6. Conclusion

Big data is defined as very large-sized, various-formatted datasets and analytic methods based on engineering technology and social network services, including statistical fusion and new visualization. A narrow definition of big data emphasizes data source, collection, storage and other technical issues, but its wider definition embraces analysis and demonstration aspects.

Among big data' data processing, visualization is a process that analyzed datasets are expressed with graph or table format. Big data's advantage in visualization in comparison with traditional data visualization is that the former uses word/text/tag clouds, network diagrams, parallel coordinates, tree mapping, cone trees, and semantic networks [Miller] more often than the latter because its data source format and their needs. R programming, Tableau software, and Python language are getting a new attention as effective visualization tool for big data demonstration.

GIS data visualization displays the spatial patterns or relationship between or among locations. Popular open source software included here are ArcGIS, Tableau, InstantAtlas, QGIS, SAGA GIS, GeoDa, and MapWindow. These tools are actively adapted to big data based software or systems to build up location oriented systems as well as more persuasive graphic works.

Big data visualization opens a new horizon in GIS visualization because it does not just strengthen the spatial context, but also it gives new meanings and insights to GIS maps and demonstration. More big data visualization skills and their outcomes will be brought out with more abundant insights and implications to GIS visualization. Especially, big data visualization can be a good measure if people involved are deliberately designed, called, instructed, and allocated.

Acknowledgements

I am indebted to Myongji University for its generous research fund in 2014. This work was supported by 2014 Research Fund of Myongji University, Seoul, Korea.

Conflict of interest

No potential conflict of interest was reported by the author.

Author details

Junghoon Ki
Myongji University, Seoul, South Korea

*Address all correspondence to: johnki70@gmail.com

References

[1] Ki J. A big data analysis of urban statistics expression—Chernoff face-based expression of local community health index in Korea. Space and Environment. 2016;**26**(1):336-358

[2] Sui D. The 'G' in GIS: Big data in a small and divided world: Implications for GIS and geography. GEO world. Business Insights: Global. 2012;**25**(3):12-13

[3] Sanderson M. Big data: GIS at the crossroads. GEO: Connexion. Business Insights: Essentials. 2018;**12**(1):6

[4] Jeong J. Three Major Factors for a Successful Big Data Application. Seoul: National Information Society Agency; 2012

[5] Data Warehousing with Apache Hadoop [Internet]. 2018. Available from: http://www.azure.microsoft.com [Accessed: Sep 1, 2018]

[6] Song K. Understanding Society through SOCIALmetrics. Seoul: Daum Soft; 2011

[7] Gao X, Cai J. Optimization analysis of urban function regional planning based on big data and GIS technology. Technical Bulletin. 2017;**55**(11):344-351

[8] Miller JD. Big Data Visualization. Birmingham: Packt Publishing; 2017

[9] Warden P. Big Data Glossary. Sebastopol: O'Reilly Media; 2011

[10] GIS Data in the Geo-database [Internet]. 2018. Available from: http://www.esri.com [Accessed: Sep 1, 2018]

[11] GIS Data Analysis Toolbox Example [Internet]. 2018. Available from: http://www.pro.arcgis. com [Accessed: Sep 1, 2018]

[12] The Flow of Human Migration [Internet]. 2018. Available from: http://www.public.tableau.com [Accessed: Sep 1, 2018]

[13] Huffman, D. On the abuse of Chernoff faces. Cartastrophe [Internet]. 2010. Available from http://cartastrophe.wordpress.com/2010/06/16/on-the-abuse-of-chernoff-faces [Accessed: Sep 2, 2018]

[14] Hesse BW, Moser RP, Riley WT. From big data to knowledge in the social sciences. The Annals of the American Academy. 2015;**659**:16-32

Chapter 8

Google Earth Augmented for Earthwork Construction Planning

Duanshun Li, Ming Lu and Rod Wales

Abstract

This chapter introduces *GoogleEarthWork* which is an augmented geographic information system (GIS) based on Google Earth to manage and visualize heterogeneous site information, especially 3D models, aerial and ground images, panoramas, and GIS data of the site environment. The concept is to realize a highly automated end-to-end earthwork construction planning system that is able to generate project management deliverables from heterogeneous information and enhance the usefulness and intelligence of GIS for better project planning and control in earthwork construction. With identified constraints from the augmented Google Earth, the earthwork planning problem is formulated, and an optimized executable plan can be automatically generated, including work breakdown structure and project network model. Demonstration cases are provided to prove concepts of and illustrate functionalities of *GoogleEarthWork* in support of earthwork construction planning in realistic settings.

Keywords: Google Earth, Keyhole Markup Language, earthwork construction, automated planning

1. Introduction

Construction project planning and control requires an integral and comprehensive understanding of the construction site. During the planning process, a large volume of data are collected and created to identify potential problems on the construction site and select proper construction methods and procedures in order to ensure safety and on-time delivery of the project. Such data include (1) as-planned information that describes the design and the scope of the project, (2) as-built information that describes the actual situation on the construction site, and (3) environmental information that can be used to evaluate the impact of the environment on the project and the impact of the project on the environment. At present, engineers and project managers can be overwhelmed with various information coming from different sources (as listed in **Table 1**); however, maintaining large-volume heterogeneous datasets would become a big burden unless they can be linked and managed together to enable efficient information retrieval and facilitate problem identification [1].

The adoption of advanced sensing and information management technologies in construction is greatly hindered by (1) high expenses on system development yet unclear benefits of implementation [20–22], (2) inefficient visualization and oversimplified site modeling methods for coping with complicated site environment [20], (3) insufficient integration and interoperability [23, 24], and (4) technology barriers and organizational difficulties in information sharing and distribution [20, 25].

Data	Usage
2D drawings (as-designed)	Design
3D models (as-designed)	Design/construction prototyping [2, 3] Site layout planning [4] Crane path and lift planning [5]
Images/videos (as-built)	Site inspection and reporting As-built modeling [6, 7] Progress monitoring [8]
Laser scanning (as-built)	As-built modeling [9–11] Progress monitoring [8]
Satellite images, topographic data, et al. in GIS (environmental)	Site layout planning [12–17] Route planning [15] Data management and visualization [1, 18, 19]

Table 1.
Typical datasets available on a construction project.

Several technologies have been applied on project information management and visualization, including building information modeling (BIM) [2], augmented reality (AR) [26–28], the integration of BIM and AR, the integration of GIS and BIM, and Google Earth, as listed in **Table 2**.

BIM demonstrates great potential to model rich geometric and semantic information of a building object but lacks the capability to incorporate as-built and environmental information. AR has gained substantial attention lately due to its capability to combine site photos and as-planned 3D models. However, the absence of an accurate model of the surrounding environment, for example, those 3D site models generally provided by 3D GIS systems, makes AR less instrumental in construction engineering applications that demand the representation of frequent, intensive interactions and relationships between the facilities being built and the site environment, especially where the project is situated in crowded cities or environmentally fragile areas. Researchers have also leveraged on the benefits of integrating BIM and AR [29–32]. Nonetheless, incorporating AR into BIM software is still practically infeasible due to inherent limitations of BIM software in handling large external datasets for real-time rendering [31].

GIS has achieved significant success in managing large-scale heterogeneous spatial information. Considerable attention has been placed on the integration of BIM models and GIS so as to integrate the indoor as-built information and the outdoor environmental information [33, 34]. To tackle unstructured data, researchers utilized variants of Extensible Markup Language (XML) to develop shared project information models thanks to its extensibility and interoperability on the web schemas [35–37]. Both the open source BIM standard of industrial foundation class

Technology	Information
BIM	3D models
AR	3D models + images
BIM + AR	3D models + images
GIS + BIM	3D models + satellite images + topographic
Google Earth	3D models + images + satellite images + topographic

Table 2.
Site information management methods.

(IFC) [38–40] and Web GIS formats (including LandXML [39, 41], City Geography Markup Language (CityGML) [42–45], and *Keyhole Markup Language* (KML) [1]) are based on XML. LandXML is mainly intended for enhancing interoperability of data utilization in the land development industry. The integration between IFC and CityGML is the most investigated approach for integrated information modeling of buildings [43]. Majority of the works have attempted to covert semantically rich IFC models to CityGML models by taking advantage of the capability of GIS to handle huge datasets with a server-based approach [34, 43]. Earlier works [43, 44, 46] in this area focused on the conversion of geometric models. Ensuing research endeavors were intended to improve the conversion of semantic information [45, 47] using semantic mapping [42, 47–49] and ontology [42, 48].

The integration of BIM and CityGML provides an effective means to manage indoor building information and outdoor environmental information. However, it lacks the functionality to support AR modeling based on site photos or videos. In contrast, KML—which represents a markup language specialized for data modeling in Google Earth—focuses on data integration and visualization. It provides various data models to support advanced visualization techniques including AR. In [50], KML was used to visualize building energy simulation results integrated with BIM. Another related endeavor [1] proposed the use of KML and Google Earth to generate a cost-effective site information management platform which integrated site photos, 3D models, and the building environment.

In this chapter, we introduce an augmented GIS system called *GoogleEarthWork*—which is conceptualized from an academia-industry joint research endeavor and prototyped by taking advantage of KML and Google Earth for managing and visualizing heterogeneous site information in support of proactive project planning and control in the particular application context of rough grading earthwork construction. *GoogleEarthWork* focuses on the integration of 3D models, aerial and ground images, panoramas, and GIS data of the site environment that are commonly used for earthwork construction planning. Such datasets are seamlessly synthesized to facilitate the identification of quantitative and qualitative constraints in earthwork construction planning through applying computer vision techniques. Further, *GoogleEarthWork* runs on an automated earthwork planner engine program, leading to the generation of an optimized earthwork execution plan.

2. GIS-based site information management and visualization

Google Earth has been widely used by scientists and relevant stakeholders in addressing environmental and construction planning issues thanks to its ubiquity and rich geographic information. Diversified geographical information is presented to the user through a combination of digital elevation models, satellite imagery, 3D building models, street views, and user-uploaded images. Features such as tiling and level of detail (LOD) for images and 3D models enable Google Earth to manage large datasets with ease and efficiency, eclipsing majority of BIM software. Besides, KML enriches the extensibility of Google Earth significantly by providing users a standardized language to add data and customize analyses. With temporal and spatial information associated with each object, Google Earth enables efficient information retrieval through content navigation, 3D exploration, and time window filtering.

The *GoogleEarthWork*—which is prototyped based on Google Earth using KML—seamlessly integrates information contained in unordered images, geometric models, and 3D GIS system. As presented in **Figure 1**, the system encompasses data collection, data processing, data management, and information visualization and distribution. Aerial and ground imageries of the construction site captured

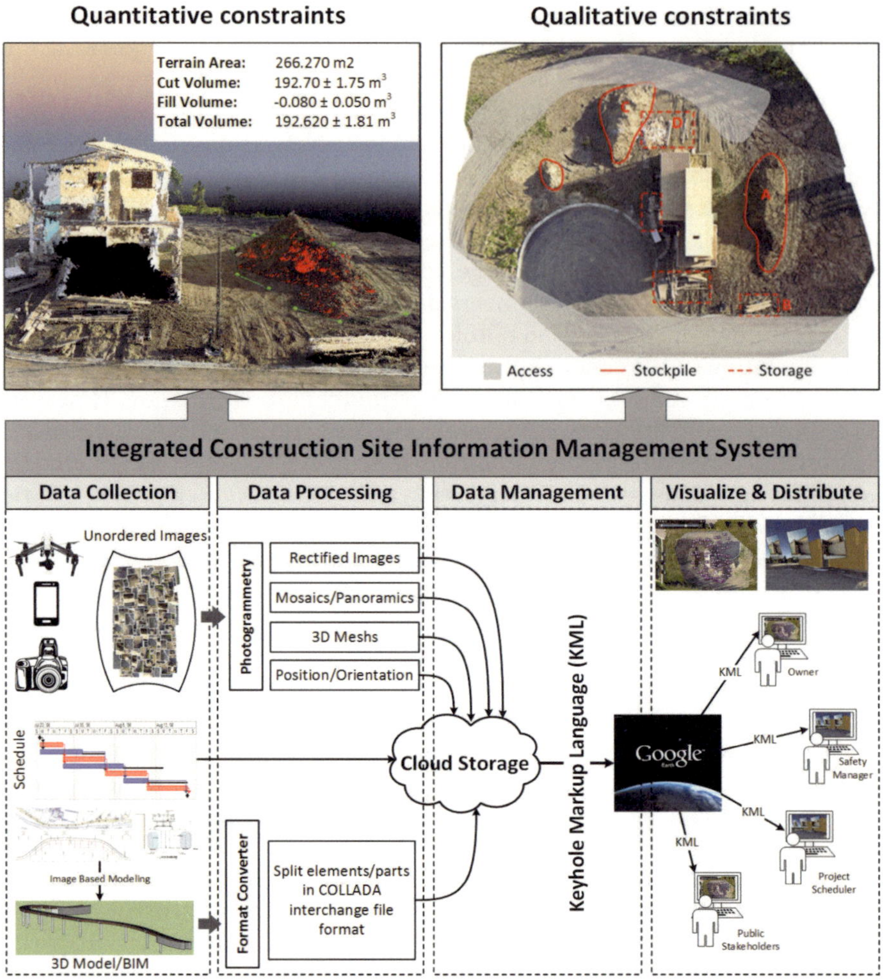

Figure 1.
GoogleEarthWork for earthwork construction planning.

with unmanned aerial vehicles (UAVs) and mobile devices are selected as major data sources for actual site monitoring and modeling. As-planned models and project schedules provide main data sources to build a virtual construction process. In addition, the 3D environment construction environment is reproduced in the visualization system. Subsequent to data acquisition, images and models need to be processed such that they are compatible with KML. Models are divided into parts in order to denote distinct construction stages in line with construction schedule.

Photogrammetry algorithms are also used in order to align unordered images within the WGS84 coordinate system adopted in Google Earth. Panoramic views and 3D reconstruction of the construction site are produced to facilitate a better comprehension of the construction environment. The resulting 3D point cloud captures the geometry of the construction site and is thereby used for cut/fill volume takeoff, as well as measuring the hauling distance between two areas. As-planned models are converted in the KML format and time-stamped in order to visualize the construction progress. The system provides stakeholders with a visually intuitive platform to perceive the construction site and identify potential problems such as spatial limits in connection with site accesses and site layouts through integrated information visualization. By storing data on the cloud, KML enables efficient management of large volumes of data in images and models. Sharing KML documents of limited size instead of original datasets also streamlines information distribution and improves computing efficiency performances.

2.1 Data collection and preprocessing

Google Earth provides project managers with free high-resolution satellite images and topographic information of the environment around a construction site. Such information is essential to plan for site accesses, site layouts, and traffic flows. As-planned information in 2D/3D drawings is crucial for scope definition, quantity takeoff, and progress monitoring. For earthwork projects specifically, the as-designed surface is required to take off cut/fill volumes. Besides, the structures being built also affect site accessibility and traffic flows.

For as-built information, site photos have been widely used on a construction site for updating construction progress and reporting safety issues or other problems. However, images collected by different personnel are barely reused due to lack of efficient image management tools. It is desirable to automatically organize images with locations in a GIS system, but the positioning accuracy of mobile devices is inadequate for two main reasons, namely, (1) low-end localization sensors embedded in mobile devices and (2) multipath effect of radio frequency signals. In general, the camera pose obtained from a consumer-grade mobile device does not satisfy the need for geo-referencing and AR applications. Higher positioning accuracy can be obtained from aerial images taken by UAV due to high-grade localization sensors embedded and lessened multipath effects. After bundle adjustment [51], the camera pose can be further improved. By taking the optimized geo-location of aerial images as references, ground imageries can also be precisely aligned in the physical coordinate system. In addition, 3D reconstruction from images is instrumental in quantifying cut/fill volumes of earthmoving jobs and fixing distances and slopes of haul roads in earthwork construction planning. Most recent research endeavors [52] have demonstrated the cost-effectiveness of UAV photogrammetry for earthwork volume estimation.

Structure from motion (SfM) [53] has been well studied in photogrammetry and computer vision domains to reconstruct the 3D structure of the scene from image collections and to recover the pose of these images. Taking unordered images as inputs, SfM outputs the precise image position and orientation, plus 3D reconstruction of the site as point cloud or model. Besides, high-resolution panoramas stitched from aerial photos are cost-effective substitutes for outdated low-resolution satellite images. As an incremental approach, SfM is suitable for processing construction site photos collected on an irregular basis along the time line. However, it requires redundant images in order to ensure "realism" of the scene. This is usually not assured when ground photos are taken by different personnel on a construction site. Therefore aerial images taken by UAV are used to materialize connecting and aligning scattered ground images. With a sequence of imageries taken on the construction site, the system implements the SfM procedure, starting from the first aerial imagery and taking it as the reference in subsequent processing of images taken by cell phones on the ground.

The direct output of SfM includes the camera pose and a 3D point cloud of the object. A much denser 3D reconstruction of the object can be achieved using stereo matching subject to coplanar constraints [54]. To visualize the 3D reconstruction in *GoogleEarthWork*, a mesh model of the object is also produced. Further, panoramic images are generated by projecting original aerial photos onto the mesh model. An example is given in the subsequent section.

2.2 Information integration with KML

Based on XML, KML uses a tag-based structure with nested elements to manage data and information associated with an object in a hierarchical manner. Different from CityGML which is designed to represent geometric objects, the strength of KML lies in visualization on a web-based GIS platform. It defines basic elements to represent geometric objects, raster images, as well as their visual effects. Elements predefined in KML are divided into several categories according to their functionality: *Feature* for vector and raster geo-data, *Geometry* for 3D objects, *AbstractView* for navigation, *TimePrimitive* for date and time, and others for visualization style, LOD, and so on. As for GIS, geo-referencing elements are the most important for defining one object. Each object needs to be geo-referenced by <*Location* > and < *Orientation* > elements. A < *Scaling* > element is also available if scaling is necessary. The detailed information can be found through the KML reference; those elements intensively used in this research are listed in **Table 3**.

Objects defined with elements in the *Feature* category are listed on the navigation panel of the Google Earth interface for interactive selection. These elements include <*GroundOverlay*> and <*PhotoOverlay*> for images, as well as <*Placemark*>, <*NetworkLink*> for geometries and models. <*GroundOverlay*> elements are used to align satellite images or panoramic images over the 3D terrain model. <*PhotoOverlay*> elements are capable to align normal images with the 3D environment for AR visualization. A 3D model can be placed under <*Placemark*> or <*NetworkLink*> elements. Geometric objects can be represented either with primary basic shapes predefined in KML or hyperlinks of models in KML files or XML-based COLLADA files [55]. <*Folder*> and <*Document*> are elements that can be used repetitively to efficiently organize hierarchical contents.

Aerial images (which are taken by UAV) provide a unique view angle of the construction site with fewer obstacles. Besides, these images can be taken on a periodical basis to capture updates and progress on site. The stitched panoramic image has much higher resolution than satellite images available in Google Earth. The <*GroundOverlay*> element can be applied to replace the outdated lower resolution satellite image with high-resolution mosaics made of most recent images. To support real-time visualization, large panoramic images are preprocessed and managed with special elements designated for visualization in different levels of detail.

On a construction site, ground imageries are usually taken at "random" locations and angles. Consequently, they are fragmented in nature and only used as evidence shown in documents in practice. However, by aligning the image at the

Element	Function	Objects
<Model>	3D model representation and visualization	3D models
<GroundOverlay>	Raster data alignment and overlay on Google Earth terrain	Panoramic mosaics
<PhotoOverlay>	Image placement and orientation for AR visualization	Original images
<Camera>	Camera position and orientation for AR and navigation	Image pose
<TimeStamp> <TimeSpan>	Associate date/time for 4D exploration of objects and activities	Schedule
<ExtendedData>	Customized data organization and visualization	Documents, webs, et al.

Table 3.
Intensively used KML elements in GoogleEarthWork.

exact location and orientation in relation to 3D models and the site environment, fragmented information provided by individual images can be well organized and seamlessly integrated. Different from real-time AR technologies which demand considerable computing resources and remain too expensive to implement on site, the *<PhotoOverlay>* element in KML affords a pragmatic approach for realizing cost-effective AR experiences and efficient site photo management. Each *<PhotoOverlay>* object is defined by (1) a *<Camera>* element specifying the position and orientation of the image, (2) a *<ViewVolume>* specifying the field of view (FOV) of the image, (3) a *<Icon>* element to store the link to the image, and (4) an optional *<TimeStamp>* element stating the date when the image is captured. Given the rotation angles of the camera (omega, phi, and kappa) obtained from photogrammetry software, the heading, tilt, and roll can be derived with equations presented in [1]. The view volume of the image can also be derived from the estimated focal length and the image size. The image capturing date and time can be readily extracted from the header of the image file; thereby, a time stamp can be added to each image to show actual progress. This also enables retrieval and viewing of images only relevant to a particular time frame.

An example of information integration in *GoogleEarthWork* through using KML is presented in **Figure 2**; ground images captured with cell phones and digital cameras, aerial images collected using UAVs, and 3D models are embedded in the Google Earth platform so that the surrounding environment of the construction site is also rendered in a cost-effective fashion.

2.3 Constraints identification in earthwork planning

Analytical simulation or optimization for construction operations planning requires knowledge of practical constraints on the construction site so as to make a sufficient problem definition. In rough grading, a certain volume of earth needs to be excavated at one area and filled at another. Accessibility issues during project execution become the primary concern for earthwork construction

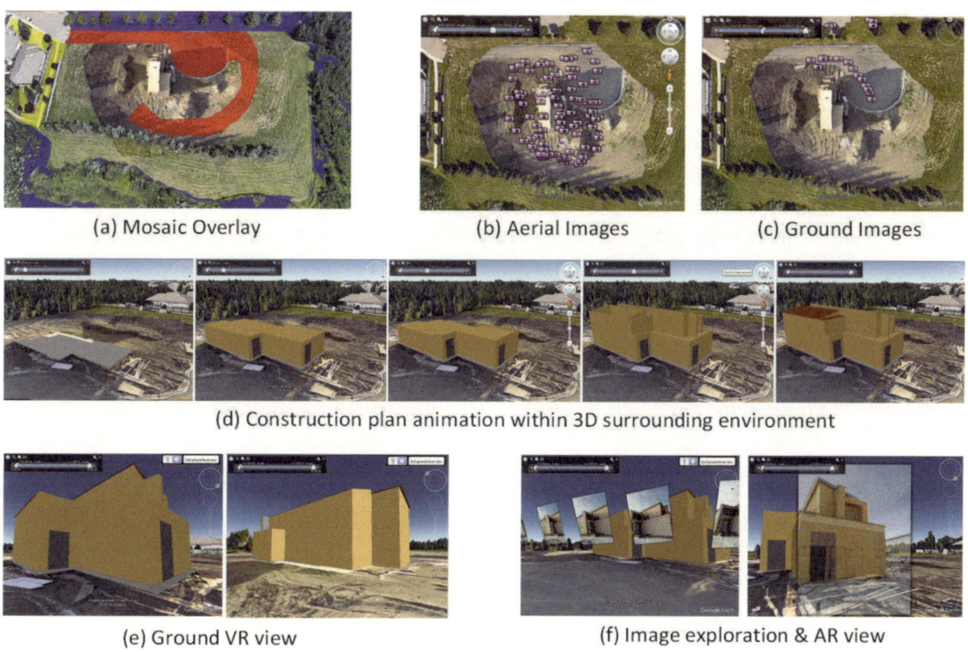

Figure 2.
Integrated project data management and visualization.

planners, especially when only limited accesses between site areas are available at the very beginning of the project. Moreover, earthmoving operations need to be executed in a safe, efficient manner, accommodating many concurring construction activities on site.

These site constraints can be categorized into quantitative constraints and qualitative constraints, as listed in **Table 4**. A quantitative constraint can be defined with a number; by contrast, qualitative constraints cannot be quantitatively represented in *GoogleEarthWork*. The two basic constraints in earthwork construction planning are (1) cut/fill volume takeoff and assignment and (2) site accessibility and haul path planning. Besides, in order to improve project performance in terms of cost and duration, a solid plan needs to consider more factors. For instance, swell/shrinkage factors account for earth volume changes during excavation and compaction. These factors have a direct impact on quantity takeoff. The haul distance, road surface condition, and slope impact earthmoving productivity. Site layout design and concurring construction activities also potentially introduce spatial constraints. For example, certain areas on site are reserved for trenching and utility installation, hence remain temporarily unpassable to trucks in earthmoving.

GoogleEarthWork assists project planners in identifying abovementioned constraints more efficiently through information integration and visualization. Among them, cut/fill volumes can be readily acquired from a dense 3D reconstruction of the construction site. The slope of the terrain can be evaluated based on the 3D reconstruction if necessary. For instance, in **Figure 3**, the volume of the stockpiles can be precisely estimated from 3D reconstruction using Pix4D as presented in **Figure 3(a)**. The relative positioning accuracy is evaluated using the width of the paved road in front of the house. The average width out of 20 measurements is 8.008 m. Detailed measurements can be found in **Figure 3(b)**. Compared with the actual width 8 m, the average error is about 8 mm, and the standard deviation is around 29 mm. Note, the absolute positioning accuracy was not evaluated in this research due to unavailability of ground truth references. Nonetheless, the visualization effect of *GoogleEarthWork* proves that positioning accuracy is sufficient and acceptable for construction planning and monitoring purposes.

Obviously, site photos provide valuable information to identify qualitative constraints. The accessibility issues, site layout constraints, and road conditions can also be assessed with high-resolution panoramic images and/or ground photos on computer. As these images are geo-located, *GoogleEarthWork* enables rapid identification of constraints at a particular spot. From **Figure 4**, it is straightforward to define one access road (pattern fill), four storage areas (solid lines), and three stockpiles (dashed lines) on the construction site directly from high-resolution panoramic image overlay.

Quantitative constraints	Qualitative constraints
• Cut/fill volume for each area	• Access to/on the site
• Soil swell/shrinkage factors	• Site layout from the design
• Traveling distance/time between areas	• Other construction activities
• Unit cost	• Road condition

Table 4.
Typical quantitative and qualitative constraints for earthwork projects.

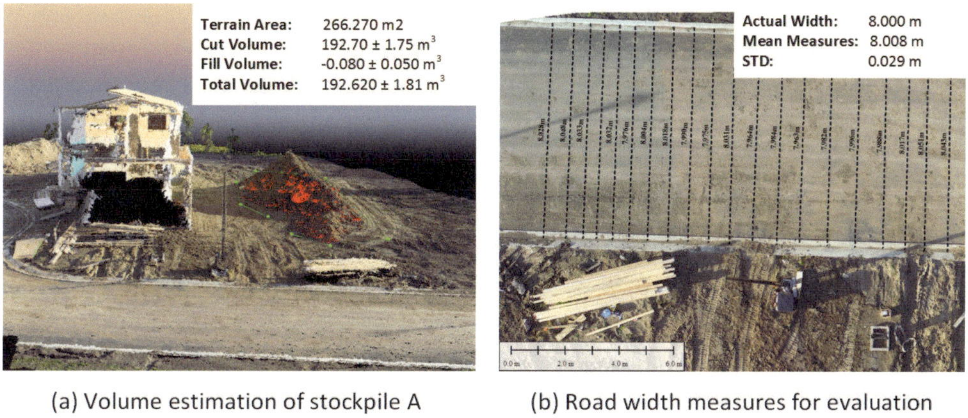

Figure 3.
GoogleEarthWork features demonstration: (a) earth volume survey from automated 3D reconstruction and (b) relative positioning accuracy evaluation.

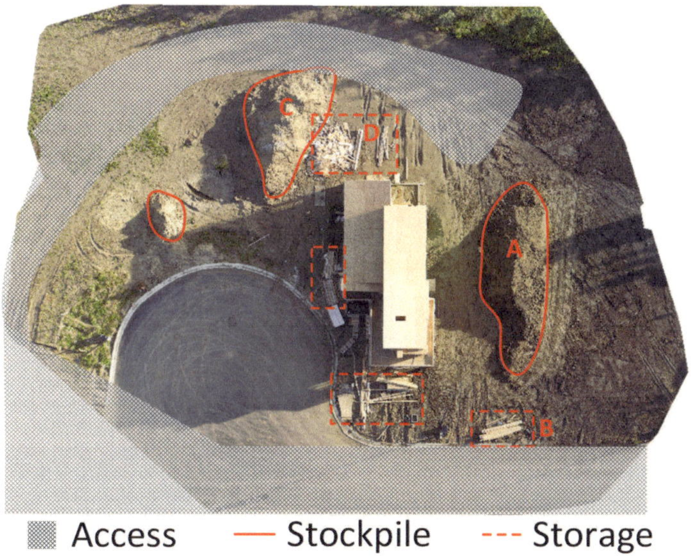

Figure 4.
Qualitative constraints identification in GoogleEarthWork: geo-relationship between accesses, stockpiles, storage area, and the building identified from high-resolution panoramic images stitched from aerial photos.

3. Earthwork optimization and planning

Given identified quantitative and qualitative constraints, the analytical method presented in [56, 57] will be introduced for automated earthwork construction planning. This method provides an analytical approach to plan rough grading operations while making problem formulation and modeling more intuitive and simplified by the use of material flow networks. To a certain extent, it can potentially eliminate temporal-spatial conflicts (such as trucks are not allowed to haul on ungraded areas) in generation of an optimized yet more practically feasible work plan. The two-phase approach splits *earthwork optimization* and *earthwork planning* into two distinct, logically connected problems. The two problems were commonly combined in previous methods; thus representing time-dependent constraints such as temporal-spatial conflicts in field operations would result in overcomplicated mathematical models which had reduced application values of the developed models in reality.

The architecture of the two-phase approach is illustrated in **Figure 5**. At the bottom, an earthwork optimizer based on a material flow network is developed to optimize earthwork operations subject to identified quantitative and qualitative constraints. The optimization result is then taken as the primary input for ensuing analysis by the earthwork planner, which generates haul jobs, defines inter-job relationships, and produces the project network model for project scheduling and control.

3.1 Earthwork optimizer

The earthwork optimizer in *GoogleEarthWork* (**Figure 6**) models site constraints and earthwork operations with a flow network model. Those quantitative constraints are defined as attributes associated with nodes, while qualitative constraints are represented in the network structure as follow: To establish such a model, the construction site is first divided into cells. For simplicity, the site can be divided into regular square cells. The links between cells can be derived directly by connecting a

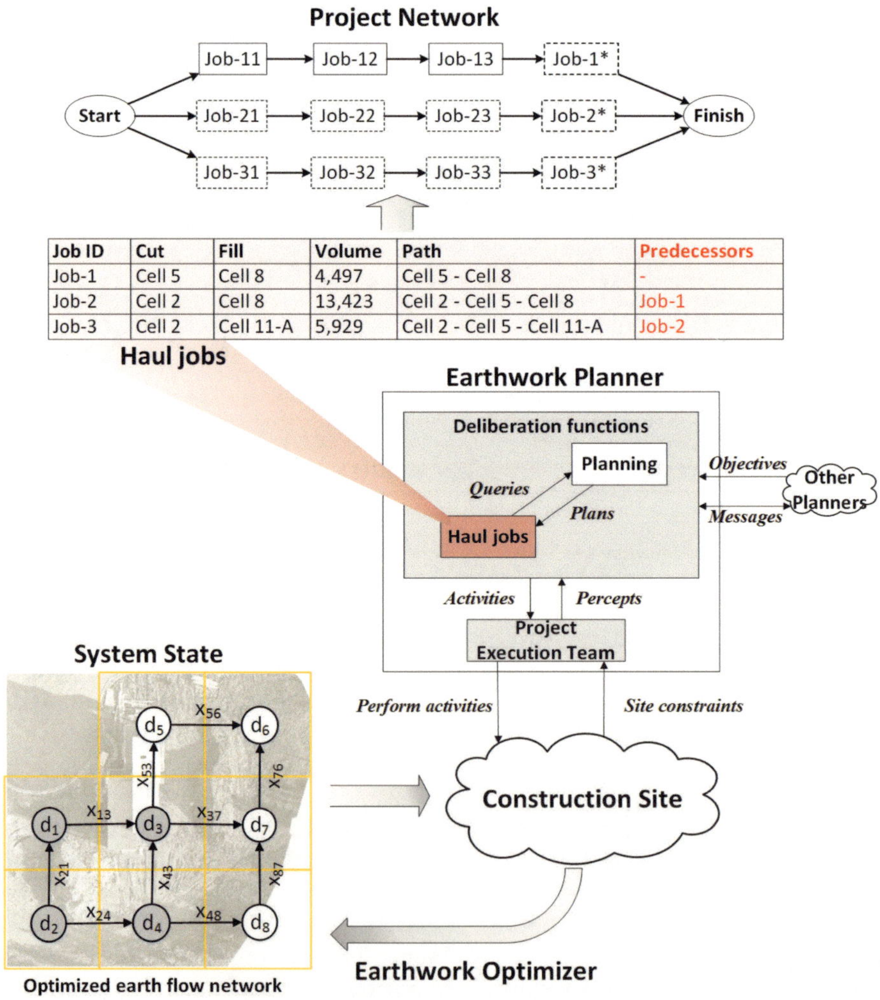

Figure 5.
Scheme for embedding automated earthwork planning system in GoogleEarthWork.

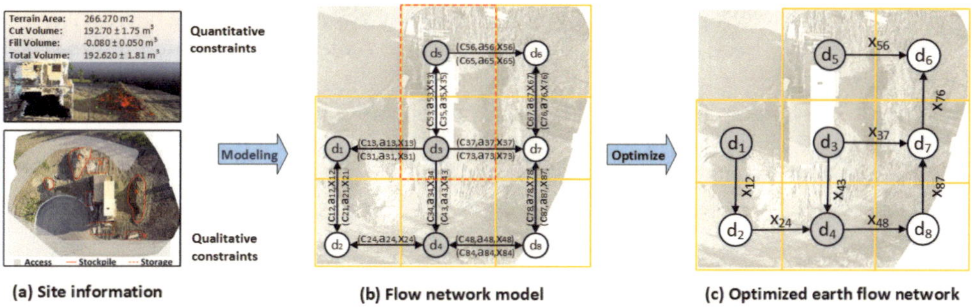

Figure 6.
Flow network model for earthwork optimization and state presentation in GoogleEarthWork. (a) Typical constraints on a construction site, (b) quantitative constraints and qualitative constraints modeled as attributes and network structure, respectively, and (c) optimized earth flow network used to represent the state of the system.

cell with its neighbor cells sharing four common edges. The division of the site into cells needs to consider the design, the site layout, elevation changes, and accesses. For example, defining separate cells at a specific position is preferred if abrupt elevation change occurs, thus resulting in definition of irregular cells. Occasionally, treating a particular site area as one node is preferable if it has limited access. In short, the identification of these problems requires integration of information on design, actual construction site, and the surrounding environment. Once the site has been digitized into a cell model, the next step is to establish the flow network model incorporating various quantitative and qualitative constraints.

Prior to delving into the core of the earthwork optimizer in *GoogleEarthWork*, several important concepts need to be clarified. A *graph model* $G = (V, E)$ is made of a set of vertices V and a set of edges E which defines the connectivity between the vertices. Given a list of vertices $V = \{v_1, v_2, \cdots, v_i, \cdots, v_j, \cdots, v_n\}$, an edge between vertex $u \in V$ and vertex $v \in V$ is defined as (u, v). For a *directed graph*, edge (u, v) and edge (v, u) represent reversed directions. A flow network is defined as follows based on the *directed graph*:

A *flow network* is a directed graph, where each vertex is assigned with a demand $d(v)$ and each edge (u, v) is assigned with a capacity $c_{uv} > 0$, a unit cost a_{uv}, and a flow x_{uv}.

The demand is the amount of flow that is required by this vertex. If $d > 0$, the vertex is demanding material to flow in. It is also called a sink node. On the contrary, it is a supplying vertex also named as source node if $d < 0$. Otherwise, the vertex will be a transshipment node with $d = 0$. The capacity c_{uv} indicates the maximum flow allowed on each edge. The cost a_{uv} is the unit cost to transport each flow unit through individual edges, respectively. The flowx_{uv} specifies the amount of flow on each edge.

The total cost of a flow network is defined as:

$$h(\mathbf{x}) = \sum_{(u,v) \in E} a_{uv} x_{uv} \tag{1}$$

where $\mathbf{x} = \{x_{uv} | (u, v) \in E\}$ represents the flow variables indicating the amount of flow on an edge. The optimal flow $\mathbf{x}_{\min}$ can be found by applying the minimum-cost flow algorithms [58] which minimize the cost function defined in Eq. (2) subject to capacity constraints and balance constraints defined in Eq. (3) and Eq. (4), respectively:

$$0 \le x_{uv} \le c_{uv} \quad \text{for all } (u, v) \in E \tag{2}$$

$$\sum_{(u,v) \in E} x_{uv} - \sum_{(v,w) \in E} x_{vw} = d_v \quad \text{for all } v \in N \tag{3}$$

Traditional methods model haul jobs directly by adding links only between cut and fill cells. These methods require predefined hauling paths which may not be explicitly specified in earthwork planning, as hauling paths can be included as variables to be optimized in addition to earth volume assignment variables between cut and fill cells. In [56], a new method is introduced to deal with the issue without increasing the complexity of problem formulation. In contrast to linking cut cells to fill cells directly, this method links neighbor cells irrespective of whether they are cut or fill cells, while the exact hauling path for each haul job will be fixed by optimization along with the source cell (cut), the destination cell (fill), and the volume to handle for each haul job.

The quantitative constraints such as cut/fill volumes and the traveling speed are directly modeled as the demand d_v for each node and the unit cost a_{uv} for each edge, respectively. The capacity of flow on an edge is typically unlimited unless there is a special need, for example, to limit the total amount moved to a storage area. The qualitative constraints are modeled implicitly in the network structure. They are embedded by adding or removing specific arcs at specific directions. In the following subsections, we will elaborate typical site constraints for earthwork including accessibility, reserved areas, and haul road conditions.

Accessibility constraints: Site accessibility constraints are the most common on a construction site. The access between cells may be blocked by waterways, ponds, other facilities, and so on. Prohibiting moving material from one cell to another may be justified in certain areas in order to ensure traffic safety and provide adequate space for other construction activities. This can be imposed by removing certain directional arcs between cells in the site grid model in order to moderate earth flows.

Reserved area constraints: Reserved areas for temporary facilities, such as fuel stations, parking yards, and rest areas, require grading as well, but trucks generally are not allowed to pass through these areas once the temporary facilities are established. They can be treated as special cases of accessibility constraints. Taking the example presented in **Figure 8**, the site is divided into regular cells. Among them, cell 3 and cell 5 cover the area where a structure is being built. After the excavation in this area, there will be a substantial elevation change. Passing through this area is thus not allowed. Considering this area is a net cut area (the total demand is negative); only material flows leaving this area (red dash rectangle) are permitted. Similarly, only material flows entering an area are allowed if this area is a net fill area.

Haul road condition constraints: It is noteworthy that the truck hauling speed on rough ground and treated ground varies significantly. In the flow network model, haul road conditions can be modeled by adjusting the unit cost a_{uv} of particular edges which represent haul road sections in the flow network. Shortening total project duration is the objective in construction planning in general. Thus, the traveling time can be used to directly model the cost.

Once the model is established, it is optimized with established minimum-cost flow algorithms [59]. As a result, the earthwork optimizer produces a flow network that defines the amount of flows (defined by $\mathbf{x}$) between adjacent cells. Because it does not model haul jobs directly, the result cannot produce the final execution plan which defines each haul job in terms of source, destination, volume, and haul path. Next, the earthwork planner is introduced which generates the final execution plan based on the optimized earth flow network.

3.2 Earthwork planner

The optimized earth flow network specifies quantity and direction to move material along inter-cell edges (haul roads) in the site system. However, temporal or spatial constraints arising from sequencing earthmoving jobs can be missed in this representation. At the beginning of earthwork operations, only limited accessibility is available. Access to an area is enabled in the middle of the earthmoving process once its neighbor areas are graded. Thus an additional network is required to define the accessibility between areas considering the progress of the project over time. In the remainder of this chapter, the optimized earth flow network is denoted as G_{op}, and the network to represent the accessibility is named as G_{ac}.

In this step, the classical planning model in automated planning theory is adopted for earthwork project planning in *GoogleEarthWork*. The state transition system for the earthwork planner is defined with a triple $\Sigma = (\mathbf{S}, \mathbf{A}, \boldsymbol{\gamma})$, where.

S is defined with a tuple of two directed graphs (G_{ac}, G_{op}), where G_{op} is the optimal earth flow network and G_{ac} is a directed graph representing the accessibility between cells.

A is the action space defined as haul jobs. Each haul job can be represented with $(WF = (Cut, Fill), P, V)$ which specifies the cut and fill cells, together with the volume V and the hauling path P. For example, a haul job $(WF = (Cut, Fill), P, V)$ indicates 20 units of material which are transported from Cell 1 to Cell 2 passing through Cell 3 and Cell 4.

y is a map from $S \times A$ to S where the optimal earth flow network and the accessibility are updated after performing an action (i.e., completing a haul job.) This includes the following:

(1) Updating the volumes of each cell on G_{op}

(2) Updating the flow between adjacent cells on G_{ac}

(3) Updating accessibility on G_{ac} after some cells are graded

In the classical planning model, actions are sequentially taken by selecting an action and updating the state as presented in **Figure 7**. The procedure consists of four steps with the first three steps corresponding to deliberation functions and the

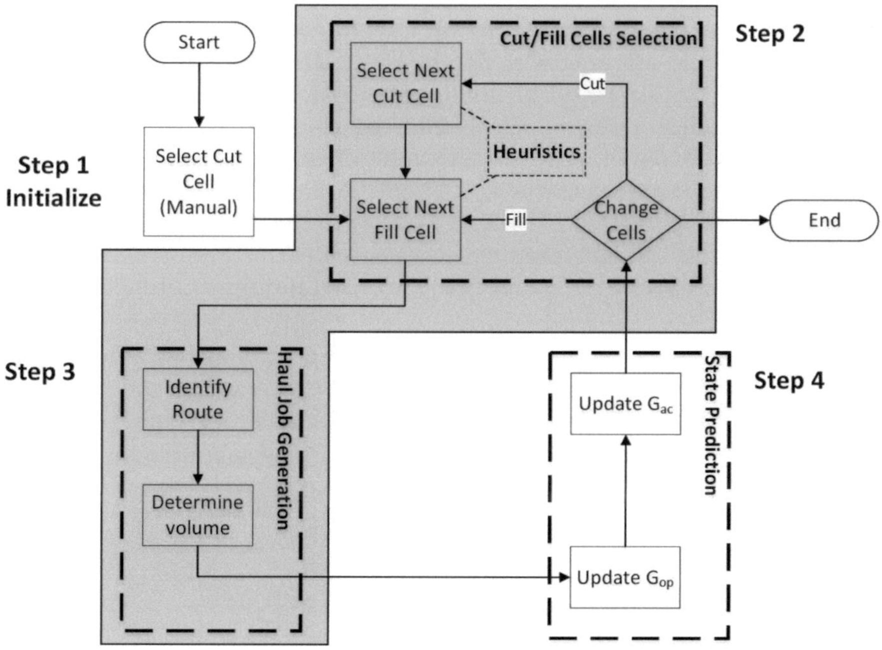

Figure 7.
The procedure of the earthwork planner. The deliberation functions include a cut/fill cell selection module and a haul job generation model.

fourth step corresponding to state transition functions. Because actions are required to satisfy all material flow constraints (flow direction and flow quantity on each edge), which are already determined in the optimized flow network, the final plan is extracted from a searching space that is already optimized. The detailed explanation of the planner can be found in [56, 57].

4. Case study

In this section a campground grading project located in Northern Alberta, Canada, is used to demonstrate the application of earthwork optimization and automated planner functions. The size of the campground is around 2000 m long and 650 m wide. The total volume of material to be handled is 584,308 bank cubic meters (*bcm*). The site layout is presented in **Figure 8** with color bands denoting deep excavation (>3 m), medium height excavation (1.5–3 m), shallow excavation (<1.5 m), shallow fill (<1.5 m), and medium-depth fill (1.5–3 m). On the west side and east side, respectively, there are two storm water storage ponds, which also provide the two primary sources for fill material in site grading. Note during construction, only limited access to the two ponds is allowed. Pond 1 has one access point on its east side; Pond 2 has two access points on its north and west sides, respectively.

A temporary haul road aligned with a future permanent road is established to facilitate the earthmoving process. Average truck speed differs when a truck hauls on the temporary road or the rough-graded ground. A fleet consisting of a 40 T excavator with a production rate of 190 *bcm* per day and CAT 740B trucks with 20 *bcm* volume capacity are employed on this project. The combined loading, dumping, and waiting time is assumed to be 20 minutes. The truck hauling speed limit, irrespective of truck haul (full) and truck return (empty), is averaged at 27 km/h on temporary haul road and 18 km/h on rough ground, respectively. Besides, hourly rates of the excavator and the truck are 140/*hr* and 135/*hr*. The hourly rate for an equipment operator is around 60/*hr* regardless of the type of the equipment.

The construction site is divided into cells (100 m × 100 m) for material flow network optimization and AON network development. The cell size is defined by the user after assessing site topology and application need. Mathematically, the smaller the cell size, the more accurate the result would be. However, too small cell size is not suitable for current application of earthwork planning and construction management. Four times the truck width is recommended as cell dimension for planning mining haul road, which was used for earthwork planning in *GoogleEarthWork* due to safety concerns [60]. To incorporate the two ponds in the flow network model, irregular shaped cells instead of squared cells are used for representing the ponds,

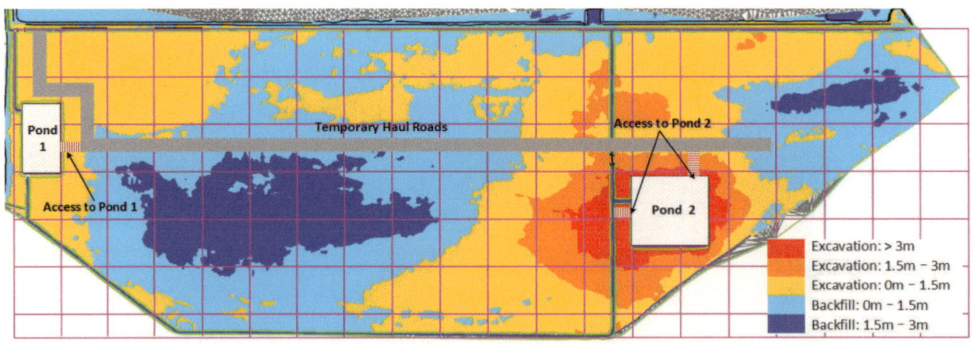

Figure 8.
Rough grading construction site: drawing and layouts.

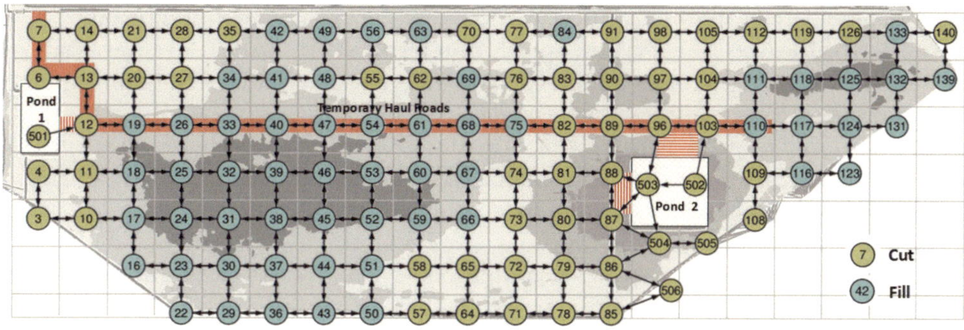

Figure 9.
Flow network model for the case study. Two ponds are split into irregular cells.

their neighboring areas, and the boundaries. In this case, Pond 1 is treated as one cell node, and Pond 2 is divided into two cell nodes. Single-directional arcs flowing out of pond cells are defined so to avoid trucks passing through the ponds. The traveling time per truckload between adjacent cells is defined as the unit cost of hauling in optimization analysis. The final flow network definition is presented in **Figure 9**.

The *earthwork Optimizer* and the *earthwork Planner* in connection with *GoogleEarthWork* were implemented based on the open source Library for Efficient Modeling and Optimization in Networks (LEMON) graph algorithm library with its LGF file format denoting flow network definition [61]. Taking the flow network model as input, an optimized earth flow network was obtained as the result of minimum-cost flow optimization. Next, eight sub-flows were identified from the optimized earth flow network based on weakly connected component analysis. In the end, a total of 129 jobs were generated. The proposed system not only enables automated project planning but also automated project network analysis and resource-loaded scheduling simulation analysis. Once the work breakdown structure and the project network are produced, they can be readily used to perform scheduling and cost analysis with existing tools. Based on the automated planner, the cost and duration were estimated to be $3,491,632 and 149 work days, respectively.

5. Conclusion

In this chapter, we conceptualize an augmented GIS system called *GoogleEarthWork* for earthwork planning based on Google Earth and demonstrate great potential in site information management and visualization, especially the integration of site photos, 3D models, and 3D surrounding environment of the construction site. The system is capable of facilitating the identification of (1) quantitative constraints by image-based 3D reconstruction and (2) qualitative constraints through interactive VR and AR visual inspection within Google Earth. Coupled with an automated earthwork planning system, *GoogleEarthWork* holds the potential to provide an integrated project planning solution that assists project managers in information collection, data analysis, and construction planning. It also enables higher-level project management analyses such as scheduling and simulation by automatically generating project execution plans (e.g., AON network model). The results provide project managers with a sufficient basis for the development of a practical, dynamic plan. As construction unfolds and the site evolves over time, additional constraints can be further imposed in order to keep the plan up to date.

At present, application of *GoogleEarthWork* is confined to rough grading earthwork construction. In the future, the system along with its underlying methodology can be customized for other application domains. More importantly, a formal definition of the *GoogleEarthWork* application framework allows the representation of domain knowledge and facilitates the implementation of advanced planning techniques (such as resource-constrained scheduling optimization or operations simulation) in the specific application domain. This will ultimately result in optimized construction plans for performance improvement in field productivity and safety. Though automated tools have been proposed for site information processing, human interactions still remain crucial in site constraint identification and planning problem definition in reality. A comprehensive information system that assists project managers in updating site constraints through site information integration and visualization is thus indispensable to refining problem definitions and identifying practical constraints. In the near future, *GoogleEarthWork* will be further augmented with advanced quantitative methods and emerging AI techniques.

Acknowledgements

This research was substantially funded by Mitacs Canada and Ledcor Canada through a Mitacs Accelerate Internship grant (Project ID: IT06594). The authors sincerely acknowledge Sam Johnson, Senior Construction Manager of Graham Construction Canada, for sharing information and insight on earthwork planning and Dr. Eric Loo, the Director of Business Development of Mitacs for facilitating the research.

Author details

Duanshun Li[1], Ming Lu[1*] and Rod Wales[2]

1 Department of Civil and Environmental Engineering, University of Alberta, Edmonton, AB, Canada

2 Ledcor Canada, Edmonton, AB, Canada

*Address all correspondence to: mlu6@ualberta.ca

References

[1] Li D, Lu M. Integrating geometric models, site images and GIS based on Google Earth and Keyhole Markup Language. Automation in Construction. 2018;**89**:317-331. DOI: 10.1016/j.autcon.2018.02.002

[2] Ahar S. Building information modeling (BIM): Trends, benefits, risks, and challenges for the AEC industry. Leadership and Management in Engineering. 2011;**11**:241-252

[3] Huang T, Kong CW, Guo HL, Baldwin A, Li H. A virtual prototyping system for simulating construction processes. Automation in Construction. 2007;**16**:576-585. DOI: 10.1016/j.autcon.2006.09.007

[4] Chang Y-C, Hung W-H, Kang S-C. A fast path planning method for single and dual crane erections. Automation in Construction. 2012;**22**:468-480. DOI: 10.1016/j.autcon.2011.11.006

[5] Tan Y, Song Y, Liu X, Wang X, Cheng JCP. A BIM-based framework for lift planning in topsides disassembly of offshore oil and gas platforms. Automation in Construction. 2017;**79**:19-30. DOI: 10.1016/j.autcon.2017.02.008

[6] Bhatla A, Choe SY, Fierro O, Leite F. Evaluation of accuracy of as-built 3D modeling from photos taken by handheld digital cameras. Automation in Construction. 2012;**28**:116-127. DOI: 10.1016/j.autcon.2012.06.003

[7] Rodriguez-Gonzalvez P, Gonzalez-Aguilera D, Lopez-Jimenez G, Picon-Cabrera I. Image-based modeling of built environment from an unmanned aerial system. Automation in Construction. 2014;**48**:44-52. DOI: 10.1016/j.autcon.2014.08.010

[8] Omar T, Nehdi ML. Data acquisition technologies for construction progress tracking. Automation in Construction. 2016;**70**:143-155. DOI: 10.1016/j.autcon.2016.06.016

[9] Golparvar-Fard M, Bohn J, Teizer J, Savarese S, Peña-Mora F. Evaluation of image-based modeling and laser scanning accuracy for emerging automated performance monitoring techniques. Automation in Construction. 2011;**20**:1143-1155. DOI: 10.1016/j.autcon.2011.04.016

[10] Turkan Y, Bosche F, Haas CT, Haas R. Automated progress tracking using 4D schedule and 3D sensing technologies. Automation in Construction. 2012;**22**:414-421. DOI: 10.1016/j.autcon.2011.10.003

[11] Zhang C, Arditi D. Automated progress control using laser scanning technology. Automation in Construction. 2013;**36**:108-116. DOI: 10.1016/j.autcon.2013.08.012

[12] Cheng MY, O'Connor JT. Site layout of construction temporary facilities using an enhanced-geographic information system (GIS). Automation in Construction. 1994;**3**:11-19. DOI: 10.1016/0926-5805(94)90028-0

[13] Cheng M-Y, Yang S-C. GIS-based cost estimates integrating with material layout planning. Journal of Construction Engineering and Management. 2001;**127**:291-299. DOI: 10.1061/(ASCE)0733-9364(2001)127:4(291)

[14] Alsaggaf A, Jrade A. A framework for an integrated BIM-GIS decision support model for site layout planning. In: International Construction Specialty Conference; Vancouver. 2017. p. 182

[15] Kang S, Seo J. GIS method for haul road layout planning in large earthmoving projects: Framework and analysis. Journal of Construction

Engineering and Management. 2013;**139**:236-246. DOI: 10.1061/(ASCE)CO.1943-7862.0000561

[16] Cheng MY, O'Connor JT. ArcSite: Enhanced GIS for construction site layout. Journal of Construction Engineering and Management. 1996;**122**:329-336. DOI: 10.1061/(ASCE)0733-9364(1996)122:4(329)

[17] Su X, Andoh AR, Cai H, Pan J, Kandil A, Said HM. GIS-based dynamic construction site material layout evaluation for building renovation projects. Automation in Construction. 2012;**27**:40-49. DOI: 10.1016/j.autcon.2012.04.007

[18] Bansal VK. Integrated CAD and GIS–based framework to support construction planning: Case study. Journal of Architectural Engineering. 2017;**23**:5017005. DOI: 10.1061/(ASCE)AE.1943-5568.0000262

[19] Oloufa AA, Eltahan AA, Papacostas CS. Integrated GIS for construction site investigation. Journal of Construction Engineering and Management. 1994;**120**:211-222

[20] Leite F, Cho Y, Behzadan AH, Lee S, Choe S, Fang Y, et al. Visualization, information modeling, and simulation: Grand challenges in the construction industry. Journal of Computing in Civil Engineering. 2016;**30**:4016035. DOI: 10.1061/(ASCE)CP.1943-5487.0000604

[21] Peansupap V, Walker DHT. Factors enabling information and communication technology diffusion and actual implementation in construction organisations. Journal of Information Technology in Construction. 2005;**10**:193-218. DOI: http://www.itcon.org/2005/14

[22] Peansupap V, Walker DHT. Information communication technology (ICT) implementation constraints: A construction industry perspective. Engineering Construction and Architectural Management. 2006;**13**:364-379. DOI: 10.1108/09699980610680171

[23] Ajam M, Alshawi M, Mezher T. Augmented process model for e-tendering: Towards integrating object models with document management systems. Automation in Construction. 2010;**19**:762-778. DOI: 10.1016/j.autcon.2010.04.001

[24] Forcada N, Casals M, Roca X, Gangolells M. Adoption of web databases for document management in SMEs of the construction sector in Spain. Automation in Construction. 2007;**16**:411-424. DOI: 10.1016/j.autcon.2006.07.011

[25] Lam PTI, Wong FWH, Tse KTC. Effectiveness of ICT for construction information exchange among multidisciplinary project teams. Journal of Computing in Civil Engineering. 2010;**24**:365-376. DOI: 10.1061/(ASCE)CP.1943-5487.0000038

[26] Chi HL, Kang SC, Wang X. Research trends and opportunities of augmented reality applications in architecture, engineering, and construction. Automation in Construction. 2013;**33**:116-122. DOI: 10.1016/j.autcon.2012.12.017

[27] Wang X, Love PED, Kim MJ, Park C-S, Sing C-P, Hou L. A conceptual framework for integrating building information modeling with augmented reality. Automation in Construction. 2013;**34**:37-44. DOI: 10.1016/j.autcon.2012.10.012

[28] Golparvar-Fard M, Pena-Mora F, Savarese S. D4AR—A 4-dimensional augmented reality model for automating construction progress monitoring data collection, processing and communication. Journal of Information

Technology in Construction. 2009;**14**:129-153. DOI: http://www.itcon.org/paper/2009/13

[29] Meža S, Turk Ž, Dolenc M. Component based engineering of a mobile BIM-based augmented reality system. Automation in Construction. 2014;**42**:1-12. DOI: 10.1016/j.autcon.2014.02.011

[30] Park CS, Lee DY, Kwon OS, Wang X. A framework for proactive construction defect management using BIM, augmented reality and ontology-based data collection template. Automation in Construction. 2013;**33**:61-71. DOI: 10.1016/j.autcon.2012.09.010

[31] Wang X, Kim MJ, Love PED, Kang S. Augmented reality in built environment: Classification and implications for future research. Automation in Construction. 2013;**32**: 1-13. DOI: 10.1016/j.autcon.2012.11.021

[32] Wang X, Truijens M, Hou L, Wang Y, Zhou Y. Integrating augmented reality with building information modeling: Onsite construction process controlling for liquefied natural gas industry. Automation in Construction. 2014;**40**:96-105. DOI: 10.1016/j.autcon.2013.12.003

[33] Zlatanova S, Stoter J, Isikdag U. Standards for exchange and storage of 3D information: Challenges and opportunities for emergency response. In: 4th International Conference on Cartography & GIS, vol. 2. 2012. pp. 17-28

[34] Fosu R, Suprabhas K, Rathore Z, Cory C. Integration of building information modeling (BIM) and Geographic Information Systems (GIS)—A literature review and future needs. In: The 32nd CIB W78 Conference. 2015. pp. 196-204

[35] Ma Z, Li H, Shen QP, Yang J. Using XML to support information exchange in construction projects. Automation in Construction. 2004;**13**:629-637. DOI: 10.1016/j.autcon.2004.04.010

[36] Song Y, Clayton MJ, Johnson RE. Anticipating reuse: Documenting buildings for operations using web technology. Automation in Construction. 2002;**11**:185-197. DOI: 10.1016/S0926-5805(00)00097-2

[37] Zhu Y, Issa RRA. Viewer controllable visualization for construction document processing. Automation in Construction. 2003;**12**:255-269. DOI: 10.1016/S0926-5805(02)00089-4

[38] Afsari K, Eastman CM, Castro-Lacouture D. JavaScript Object Notation (JSON) data serialization for IFC schema in web-based BIM data exchange. Automation in Construction. 2017;**77**:24-51. DOI: 10.1016/j.autcon.2017.01.011

[39] Kang TW. Object composite query method using IFC and LandXML based on BIM linkage model. Automation in Construction. 2017;**76**:14-23. DOI: 10.1016/j.autcon.2017.01.008

[40] Redmond A, Hore A, Alshawi M, West R. Exploring how information exchanges can be enhanced through Cloud BIM. Automation in Construction. 2012;**24**:175-183. DOI: 10.1016/j.autcon.2012.02.003

[41] Le T, Jeong HD. Interlinking life-cycle data spaces to support decision making in highway asset management. Automation in Construction. 2016;**64**:54-64. DOI: 10.1016/j.autcon.2015.12.016

[42] Deng Y, Cheng JCP, Anumba C. Mapping between BIM and 3D GIS in different levels of detail using schema mediation and instance comparison. Automation in Construction. 2016;**67**: 1-21. DOI: 10.1016/j.autcon.2016.03.006

[43] El-Mekawy M. Integrating BIM and GIS for 3D City Modelling: The Case of IFC and CityGML. 2010. DOI: urn:nbn:se:kth:diva-28899

[44] Hijazi I, Ehlers M, Zlatanova S, Isikdag U. IFC to CityGML transformation framework for geo-analysis: A water utility network case. In: 4th International Workshop on 3D Geo-Information. 2009. pp. 123-127. DOI: 10.13140/RG.2.1.4623.0246

[45] De Laat R, Van Berlo L. Integration of BIM and GIS: The development of the CityGML GeoBIM extension. In: Advances in 3D Geo-Information Sciences. 2011. pp. 211-225. DOI: 10.1007/978-3-642-12670-3_13

[46] Isikdag U, Zlatanova S. Towards defining a framework for automatic generation of buildings in CityGML using building information models. In: 3D Geo-Information Sciences. 2009. pp. 79-96. DOI: 10.1007/978-3-540-87395-2_6

[47] Donkers S, Ledoux H, Zhao J, Stoter J. Automatic conversion of IFC datasets to geometrically and semantically correct CityGML LOD3 buildings. Transactions in GIS. 2016;**20**:547-569. DOI: 10.1111/tgis.12162

[48] El-Mekawy M, Östman A. Semantic mapping: An ontology engineering method for integrating building models in IFC and CITYGML. In: Proceedings of the 3rd ISDE Digital Earth Summit. 2010. pp. 1-11. Available from: http://news.digitalearth-isde.org/Bulgaria/pdf/32_El-Mekawy_Sweden_paper.pdf

[49] Cheng JCP, Deng Y, Anumba C. Mapping schema semi-automatically utilizing linguistic and text mining techniques. Journal of Information Technology in Construction. 2015;**20**:193-212

[50] Niu S, Pan W, Zhao Y. A BIM-GIS integrated web-based visualization system for low energy building design. Procedia Engineering. 2015;**121**:2184-2192. DOI: 10.1016/j.proeng.2015.09.091

[51] Triggs B, McLauchlan PF, Hartley RI, Fitzgibbon AW. Bundle adjustment—A modern synthesis. Vision Algorithms: Theory and Practice. 2000;**1883**:298-372. DOI: 10.1007/3-540-44480-7_21

[52] Siebert S, Teizer J. Mobile 3D mapping for surveying earthwork projects using an Unmanned Aerial Vehicle (UAV) system. Automation in Construction. 2014;**41**:1-14. DOI: 10.1016/j.autcon.2014.01.004

[53] Snavely N, Seitz SM, Szeliski R. Modeling the world from Internet photo collections. International Journal of Computer Vision. 2008;**80**:189-210. DOI: 10.1007/s11263-007-0107-3

[54] Furukawa Y, Ponce J. Accurate, dense, and robust multi-view stereopsis. 2008;**1**:1-14. Available from: http://www-cvr.ai.uiuc.edu/ponce_grp/publication/paper/cvpr07a.pdf

[55] COLLADA. 2017. Available from: https://www.khronos.org/collada/ [Accessed: April 3, 2017]

[56] Li D, Lu M. Automated Generation of work breakdown structure and project network model for earthwork project planning: A flow network-based optimization approach. Journal of Construction Engineering and Management. 2016;**143**:4016086. DOI: 10.1061/(ASCE)CO.1943-7862.0001214

[57] Li D, Lu M. Classical planning model-based approach to automating construction planning on earthwork projects. Computer-Aided Civil and Infrastructure Engineering. 2018;**00**: 1-17. DOI: 10.1111/mice.12420

[58] Goldberg AV. An efficient implementation of a scaling minimum-cost flow algorithm. Journal of

Algorithms. 1997;**22**:1-29. DOI: 10.1006/jagm.1995.0805

[59] Király Z, Kovács P. Efficient implementations of minimum-cost flow algorithms. Informatica. 2012;**4**:67-118. Available from: http://arxiv.org/abs/1207.6381

[60] Liu C, Lu M. Optimizing earthmoving job planning based on evaluation of temporary haul road networks design for mass earthwork projects. Journal of Construction Engineering and Management. 2014: 1-14. DOI: 10.1061/(ASCE)CO

[61] Dezso B, Jüttner A, Kovács P. LEMON—An open source C++ graph template library. Electronic Notes in Theoretical Computer Science. 2011;**264**:23-45. DOI: 10.1016/j.entcs.2011.06.003

Section 5

Volunteered Geographic Information

Chapter 9

Volunteered Geographic Information System and Its Contribution in Service Sector Employment

Nuggehalli Narayanachar Ramaprasad and Priya Narayanan

Abstract

The potential of identifying the position of the observer in the geographical system in relation with neighbouring details and the instantaneous communication to the web server has created big data in geographic information, which has evolved in web Geographical Information System (GIS). Web GIS server provides the most reliable information in real-time. The technology is in its newbie has fascinated every age group to become the member of Web GIS. The use of technology and services are so user-friendly that even a layperson can handle without errors. The technology is finding its way in many service sectors like E-Commerce, vigilance, security and real-time tracking. The scope of expansion of this technology regarding the user community and knowledge utilisation awaits its versatility. The augmentation of techniques with its unique combinations has resulted in the building blocks of innovation and citizen-friendly apps. The growing user community is, in turn, encouraging the amateurs and professional to contribute to it. The web GIS facilitated service sector employment, which provides service at doorsteps, hence creating a scope of opportunities for young entrepreneurs.

Keywords: volunteered geographic information, location-based service, global positioning system, E-Commerce, vehicle tracking

1. Introduction

The volunteered geographic information which emerged, when the high-resolution satellite images were made available. High-resolution imageries attract the public to visualise the world from the sky (bird's eye view). People are enthusiastic to visualise the earth and gradually the identification process that contributes to the improvement in clarity. The raster-based maps got digitised to vector format by the creation of features by the stakeholders. There is a long history for the human to use maps, while making different types of decisions, from daily route planning to national military strategy. Maps or the geographical information on the maps have become the obligatory part of our daily life. This paper aims to use an example and initiative of using GIS in E-Commerce application of construction and architecture. GPS and GIS systems with similar functionality will more widely be adopted or even would become a must in future E-Commerce applications. The task of creating such huge data and its updating is almost impossible for any single organisation. With the advent of web 2.0, many individual contributors simultaneously generated the digital data.

The employment of the service sector was providing some service at a particular location at a specific time. The service type may be on a regular basis or a temporary basis. The regular service is a traditional service in the regular course of time like the supply of milk and newspaper, school bus and postal services and the like. The occasional services like the personal transport, delivery of goods, painting and catering, housekeeping and maintenance and building construction require an exact location for prompt delivery.

Moreover, some of these services required at doorsteps, which were very difficult for strangers to deliver/provide service in a particular geographical location because of the lack of the locational knowledge about the area. The ability to provide the locational information through web-based GIS servers on the go, in the smartphones as shown in **Figure 1** improved the services from providing agency to the consumers.

The added advantage of the voice-based response assisted the drivers during navigation. The safe driving improved, and the advanced information regarding route and driving instructions of the navigation with the real-time traffic and weather information and the distance to be covered with the travel time ensure the punctual destination reach. Advancement in the domain network and computing has contributed to the evolution of the Spatial Data Infrastructure (SDI) enterprise. Targeted at providing access to the information that they require and an environment in which geospatial users and producers contribute, SDIs are still in the progressive stage, but the architecture is a necessary and useful mechanism for LBS. Wireless data communication combines the mobile and spatial data communication by giving consumers easy access through wireless devices (mobile telephones, personal digital assistants, palmtop computers) to relevant information on the Internet and intranets. The consequential technology is the positioning technology, the way to find out the position of a mobile device accurately. Referable to the unique

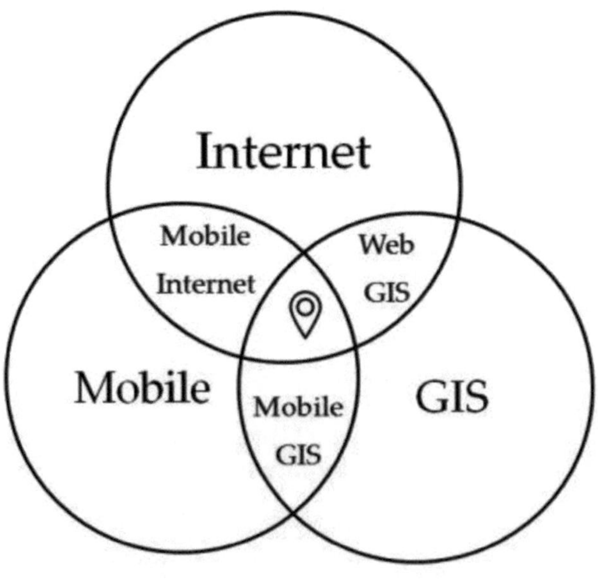

Figure 1.
Structure of smartphone-based geographical locational service [1].

features of the cellular environment, it is a significant challenge to locate the target precisely. E.g., global positioning system (GPS), assisted GPS, network-based technology, the travel time of signals.

2. Crowdsourcing

Crowdsourcing is the concept of integrating the work of the significant participants (crowd). Here the smallest contributions are accumulated to do large tasks which are having hardships physically and economically. It is the more public contribution of work than of the specific professional task, most commonly operated and owned by non-profit making organisations having public involvement. Before the digital era, the term crowdsourcing referred for creating the funding procedure and creating commons of public interest like building communal spaces like the water bodies, parks and vacation spots. In the digital era, the concept of crowdsourcing refers to the creation of data, computation and analysis. It also channels the experts who wish to solve an issue and share the solution to the general public [2].

3. Volunteered geographic information

Virtual globes have unlocked the world of satellite imagery to the general public. They allowed education, entertainment, and exploration for the newbie's [3]. Importantly, a domain to communicate their investigations and information that the public is interested in [4, 5]. The development of Digital Earth (DE) is focused on the whole of humankind to progress considering humanity as a whole. The determination is towards the positive contribution for public peace, safety, order, and general prosperity consistent with the Brundtland Commission on sustainable development [6]. DE is as much social science as it is a phenomenon of quantitative science and engineering. It should equally respect as the social science with an ontology and epistemology, i.e. the study of the nature of the 'reality' of DE (ontology), and the study of the process of the acquisition of knowledge [7].

Geo-informatics is an essential tool in the developing cultural universe of human society. The collection of Geospatial Information with its attributes and its regular updates is a gigantic task in the dynamically changing complex world. The increased precision of positional accuracy and availability of latest high-resolution satellite imageries is the source of Geo-information in Participatory Geographical Information System (PGIS), further enhancing this is the time-to-time updated information in the reliable real-time world. The citizens act as sensors to capture spatial data; these data are edited and managed by a collaborative environment [8, 9].

Volunteered Geographic Information (VGI) is all about the location information of mundane things appearing on the earth surface which is recorded by volunteered people through Crowdsourcing methods. The traditional techniques of geographic information were intern gathered from surveying techniques by professionals. Whereas, the conventional methods were accurate but failed to keep the map information up-to-date. The task of disintegrating the work and distributing work to many people mainly called outsourcing, in the corporate sector this system introduced for the economic purpose. In many cases, firms took risks experimenting with outsourcing while lacking a firm understanding of the relationship between internal functions and its spatial implications. Outsourcing has set companies with unexpected drawbacks. The miscommunication and lower quality of products, absence of completeness and focus, which ended in delaying of the production process. Technological advances have surpassed some of the needs business process outsourcing. The volunteered contributions cannot be pejorative, as they arrived from local expertise having a thorough knowledge of the place. The VGI is open data available for all, and volunteers contributed it. The task of identifying geographical location needed rigorous training in the field of mapping and surveying. The high-resolution satellite imageries made this task more straightforward and comfortable even for amateurs. Some of the natural colours like the blue waterbodies and green trees in the colour satellite imageries made the identification task even more straightforward. The correlation techniques with already available information in the neighbourhood facilitated them to locate the positions and explore more new things. As the process tasks within firms remain unclear, there is a degree of uncertainty about which jobs need to stay geographically clustered together.

The reduction in the monetary value of GPS receivers, with an increase in positional accuracy and the amalgamation of GPS with smartphones, made this technology affordable for the common man. The availability of high-resolution satellite imageries including spatial and attribute data like the Google Earth, Open Street Map (OSM), Bhuvan, Bing maps, in the active web-publication makes it easy to identify and connect the geospatial happenings. The identification of known familiar features in adjacency will resort for confirmation of nearby details. Demand and manufacturing in large-scale reduced hardware prices. Free and open source software's encourages investing intelligence in the geospatial world and befits reaped by all for constructive purposes.

Techniques of pathfinder for converting the crowd into the resource that can provide information through predominantly designed apps ultimately depends on the popularity, awareness, necessity and usage and literacy rate in the region. Whereas, the data collection in the peacetime corresponding to all the major theme may not be exhaustive. The evolution of virtual earth technology has provided access to low-cost and easy-to-use methods and to communicate geospatial information more effectively among the general public, as well as among scientists [3, 10]. Crowdsourced data via smartphones will generate data in real time. This information has enormous perspective in emergency service and disaster management [11]. The inception of open access science, open source software development, open data, social media, transparency and open government movements, have made public sector data accessible freely.

The achievement of long-term VGI sustainability is possible, only if citizens can contribute content to projects where 'these contributions are facilitated unobtrusively, casually, or, even calmly. That the current one-size-fits-all software interfaces for crowdsourcing spatial data may alienate new, inexperienced contributors and expert veteran contributors in equal measure [9]. Geo servers provided access to satellite images to the general public and allowed entertainment, education and exploration of new findings [3, 4]. Mobile phones with integrated with advanced navigation features' for geospatial information service's with mobile Web browsers promotes active as well as passive crowdsourcing. There is now a general acceptance from the public administrators that data are state assets to be shared freely with citizens, civil society organisations and the private sector for the developmental projects [12].

4. Comparison between open and closed VGI

The comparison between the Google Maps (closed) and OSM (open) cannot conclude which one is better, both have many similarities and differences, but their design based on different fundamentals, but both provide the geographical information. The critical difference between OSM and Google Maps are on their philosophy of "Open" vs. "Closed" approach, concerning data collection and distribution. In OSM that the user and the community own every edit user make, while every change user makes to Google Maps is owned by Google. The OSM has over 2.2 million registered users who are updating the detailed map of the entire world which makes the project successful. Every update is instantaneously visible to all other users and is version controlled. There are no corporate map release cycle and approvals that are typical to large organisations. Altruism, professional interests, intellectual stimulation, protection or enhancement of personal investment, social reward, enhanced own reputation, provides an opportunity for creative and independent self-presentation; and finally pride of place [12]. However, some negative motivations can also be seen, which could preclude the automation of contributions into a system, and these are mischief, an agenda that can bias contributions, and malice and criminal intent [12]. Long-term VGI sustainability is possible if citizens can contribute to the contents of the projects were these contributions facilitated unobtrusively, casually, or, even calmly [12, 13]. The argument that OpenStreetMap vulnerable to mapping vandalism, subsided by a rapid response of correcting errors, even closed source maps are also equally vulnerable.

5. The response of volunteered geographers during an emergency

The actively registered members of OSM mapping community are responding quickly to the situations of emergency by educating users and converting them into resources. The power of the OSM community is visible especially during the humanitarian crisis of natural disasters. When the Haiti island suffered 7.0 Mw magnitude earthquake on 12 January 2010, the OSM community within a couple of hours created the map of the whole island from satellite imagery, which helped relief workers to properly coordinate rescue missions and save many lives, while commercial maps were not able to respond in such a short duration. The more recent Nepal Earthquake on 25 April 2015 with a magnitude of 7.8 Mw witnessed over 2000 volunteers map contributions responding to the emergency within 48 hours. These maps helped rescue operators in rendering the helping hand to victims, clearing the debris and providing necessary food and shelter for the affected persons.

6. Challenges of VGI

The difficulties with crowdsourcing are the generation of interference and the redundancy of data. The crowdsourced spatial information is very economical, and at the same time, the question of completeness is also significant. The VGI created data mainly focus on places of importance, and sometimes it may be void trivial remote areas. The personalised data uploaded by volunteers are mostly of advertisement in nature instead of information. However, there is a little difference between the advertising and information contents, as distinguishing them in service matters cannot be bifurcated easily. The time-stamped (metadata) attribute data is equally essential in respect of spatial data for temporal analysis. Primary virtual globe software performs similar functionalities such as spatial querying providing satellite/aerial images, topographic maps, GPS, spatial quarry, integration, distance, area measurement, movie makers, 3D graphics and topography and Wikipedia integration [3]. Currently, the usable information will be up to street level and the data at the floor level if available through crowdsourcing makes the complete GIS of citizen charter. The spatial information can facilitate E-Governess and decision support system.

7. Why is VGI required?

Any single organisation cannot generate the significant spatial data. The developed nations are also striving hard to update the national maps saving forbidding them to be obsolete. The Volunteered Geographic Information (VGI) is becoming popular in this regard as the origin of spatial data. This technology still being in its amateur stage, refinement of the protocols of the data model structure and the standardisation is under continuous evaluation. The purpose of spatial information also invites the citizens for volunteered contribution. Introduction and encouragement for spatial data utilisation in school curriculum become a significant achievement in systematic learning. There has already been a drive for environmental awareness and its protection. They can further be augmented by smartphone-based apps to explore and protect the natural resources leading to sustainable development.

The urge for reliable information is growing day by day. No single organisation could provide the domain information at pace with the dynamics of spatial data; hence the VGI system becomes an indispensable component. The well-distributed crowd as access to all the geographical locations and chances of omission becomes remote. The web servers that provide the information instantly by crowdsourcing methods the chance of visiting all the places is high; thereby updating the maps available on the web to be the latest is the best possible solution. There is an ever-increasing demand for reliable geospatial information among the public. The expenditure of surveying and spatial data collection is becoming very expensive. The availability of map/digital data in national mapping agency is obsolete; to keep the map updated in real time the option of crowd editing seems to be the only solutions. The governmental mapping agency used to embrace the strict methods of extensive surveying and mapping with the modern engineering science of mathematical function. The update and refresh rates of data for urban areas are often rapid in OSM. However, in many developing countries The citizen-based mapping using the satellite imageries are usually far superior to the mapping products generated by the national mapping agency in those countries [9].

8. Data structure

The advantages of the on fly transformations have become accessible to mash-up spatial data from different sources and integrate them. The scale-free digital data in ground terms have the advantage of displaying on various scales.

The open street maps, wiki maps have provided a platform for volunteers to enter/edit/delete any geospatial element thus providing a broad range of tools for editing. Web 2.0 has changed the internet from a passive to an active experience where users can participate [11]. With the advent of web 2.0, the user can actively participate in the live mode and upload and download data comment share and even compose maps online.

Moreover, the beauty of the system is that multiple persons can update the map simultaneously by sitting at various locations. Some of its outstanding contributions are Crowdsourcing, collective intelligence, web service, data sovereignty and design for flexibility. The mashup can combine a map source with reliable information, and people can create a latest interactive map on the Geoweb using an Application Programming Interface (API) with little or no programming knowledge [11].

Intelligence drives the semantic web, and it delivers the information which the person needs and precisely for the observers requested location. The report of the flood, weather, and traffic information is shared instantly for the location of the user. The user likes and needs like restaurants; tourist place enroot of the journey is displayed with voice responses.

Geographical features grouped into point, line, area and volume features. Geographical names of location displayed as text features. Some of the descriptive remarks also shown as text. Some attributes highlighted as tooltips upon pointer location. The point and area features are interchangeable depending upon the scale displayed. The line features generally depict the continues features like the roads and railway and rivers. The linear features are communicative they are also used for network analysis (for finding the shortest path or shortest duration). The surface features depicted as contours and hillside. The boundary lines and contour lines are imaginary lines which will not be visible on the ground and extracted by different means. Boundary lines are surveyed by locating the existing pillars and interpolating the remaining portions from the sketch. The contours are stereo photogrammetrically plotted using overlapped aerial photos or by scanning through the RAdio Detection and Ranging (RADAR) or Light Detection and Ranging (LiDAR) sensors. The point features are a symbolic representation of features for visual communication. The area features have a commendable geographical extent. The forests, cultivable area, artificial lakes and town limits shown as area features. For visualisation purposes, the line features are classified based on styles, thickness and colour. For area features like water bodies and forest, natural colours blue and green used. The point symbols are cartographic symbols used similar to the profile and planned symbols, the plan symbols like the well and huts appear as displayed from the zenith, the profile symbols are as it appears from the observer's position like the temple and trees. More or less of the conventional symbols are utilised by the institutes like the hospital and police station are used in the maps.

The data model structure of the spatial database for national mapping depended on the publication scale of the map. The scale-free digital data, there will be ambiguity while determining features. However, the standardisation of symbols with data model structure can resolve many issues.

While displaying, it should be carefully generalised, emphasising the important while removing the unimportant, group the information both thematically and perceptually with attention to visual hierarchy. The good cartographic design principles to adhere to modern interaction design paradigms [14, 15]. Furthermore, researchers in the cartography and geo visualisation domain have taken a keen interest in cognitive, and usability issues and much progress has been made to understand how human capacity can enhance or limit our experiences with visual displays [16–20].

9. Data quality, redundancy and challenges

The topological relations are crucial for the spatial data avoiding Dangles, Overshoots, and Slivers creates an error-free spatial data. Proper snapping and registration between neighbouring features assure topologically clean data. The relations like neighbourhood adjacency should be the thought in high schools like the logical gates and regular mathematics by adding to the curriculum. However, video lectures training helps amateurs in generating a topologically clean data. Such acts of necessary learning create interests and increase volunteered contribution. The use of structured query language helps to extract the information. Some of the tools like auto snapping and continuation tools correct the errors automatically. Citizens, experts, and non-experts alike are increasingly participating in the process of creating seamless spatial information and collaborating with others management skills in problem-solving tasks [9]. We have active citizens effectively engaged in the sharing of information, amateurs preferably depict the places of public importance, where the transformation occurs. The location app on the smartphone enables them to acquire the spatial details. The VGI created by amateurs may have minor identification and positional errors, but in large they give the latest information. Upon use of GPS positional errors are negligible regarding the location-based applications. However, they can opt in a drop-down menu to choose the skill set of the user.

The most faced challenges are the redundancy of Information. The prominent, easily accessible features crowded with the multiple entries shall optimistically view as the confirmation of information and also instigate the community waking of the public towards urban ecology. Like the Wikipedia, the, if the hierarchal system of data checking and accepting entries and edits adheres, along with data validation eliminates the noise. The geospatial information, thus generated can serve the humanity. The quarry based data collection and verification from registered active volunteers near the incidence aims at the protection and building awareness by the participatory public and ascertaining through physical verification in case of violations. Web published maps are reliable; real-time updated latest information is a modern gadget in everybody hand for geospatial information.

The reluctance showed on the part of contemporary governments to accept, use and disseminate crowd-sourced data for some valid reasons, such as data accuracy, reliability, and authenticity; technological and human resource limitations; and because they do not source from 'authoritative' organisation [12, 21]. Many countries launched their national virtual globes in the web portal for national spatial data; some example is French-Geoportal (http://www.geoportail.fr/) and India-Bhuvan (http://bhuvan.nrsc.gov.in/). With the progress in science technology and data infrastructures, policies of the open government, the public sector is no longer contemplate to be the sole arbiter and producer of state knowledge, and it no longer maintains an institutional monopoly [12, 22].

10. Relative spatial data sharing from other agencies

Adopting data exchange and interoperability specifications allow the fusion of heterogeneous systems and different data in one platform. The rapidly developing study of 3D mapping and modelling of the Earth holds commitment for implementation at the planetary scale [3]. Easy-to-use and intuitive virtual globe technologies have distinct advantages over conventional GIS [3]. Designing a user-friendly application for mash-up and API for determining the violations of natural resources. Descriptive information about the migratory water folks, the growth of aquatic weeds and quality and table of underground water should be web published every year to ensure the very existence of lakes in its state of health. The dissemination of knowledge and data for the end user will create a geospatial database by volunteers that serve the entire world.

The topographical information like the hydrology, land cover and hypsography are difficult for amateurs in representing as they require skill, training and experience in depicting the data. The imaginary lines like the political/administrative/revenue boundaries are useful if appended to the crowd-sourced digital data. The kilometre stones and kilometre numbers/bridges which associates itself with corresponding details of road and railway, and canals, lamp post numbers, pillar numbers they become the identification marks in identifying the relative location of the concerned space. Well, identifiable places should be made mandatory by the concerned road and railway department to share in the public domain. It is essential to integrate real-time in situ field data, remotely sensed data and geo-information system data for processing and analysis tools into a systematic framework aiming at handling complex geospatial data at the global scale [3]. The integration of remotely sensed environmental (thermal, normalized difference vegetation index (NDVI)) data into a GIS platform can aid in a better understanding of the spatial–temporal dynamics of a wide range of Earth/ecological/disease systems, especially those with spatial/environmental correlates [3, 23].

11. Visual reporting

The geo-tagged spatial information like the pictures, videos through modern smartphones and camera becomes first information report to have the ground truth. Photos and videos are all cameras with drones are an added advantage to access the damage happened, and also they give a hint to the rescue team in assists. Digital signatures and geotagged photos confirm the prompt delivery of goods. The natural resources that are of aesthetic, cultural and economic value can also be published for its best use while giving impetus to local tourism and revenue generation.

12. Location-based service

The broad sector of services that can be thought off rely on users' location information, although the E-markets are finding their way. The main point is to remember that location is the useful bit of data that can be used to extract access to many types of geographical information and services. There are numerous ways to capitalise on location to provide more relevant, useful information, or derive new services. It can be particularly compelling when combined with other user profile to offer personalised, and location sensitive responses to customers [24, 25] differentiates emergency services, mobile operator services, and value-added services (VAS), focusing on the latter category as an opportunity in the expansion of E-Commerce.

Location-based application areas identified: communication, fleet management, routing, safety, security, and entertainment. Their services classified according to whether they apply to consumers, business customers, or employees in a firm as described in **Table 1** which are customary. Most likely, LBS can provide combinations of available revenue models. The customers may be offered the choice between the advertiser and non-advertiser-supported services, with the former provided free of cost and the latter charged for service.

Existing proposals from operators and standardisation bodies specify a priority scheme whereby the core network elements (e.g.. Home Location Registers) have master control on location information. The disclosure of such information to other agencies (e.g., location servers, LBS serving nodes, Application Service Provider (ASP) is subject to subscriber needs (e.g., registration information) and regulatory frameworks.

The heterogeneity in user service needs is likely to be a feature of location-based events. It is crucial to develop suitably flexible middleware to support application developers in a 'pick-and-mix' approach to combining devices, networks and sensors. Upon entering the business, a local Wireless Fidelity (WiFi) network may

Mapping, discovery and infotainment	Searching places and location information	Finding places, friends, maps, add location to posts.
Tracking	Tracking of the cab movement for safety (Single lady passenger). 	Vehicle & Fleet, Supply chain & Inventory
Advertising	 Amusement parks, Malls, cinema hall, restaurants 	Mobile coupons, ads, deal finders
Location Based Games		Real life games like Google Ingress, Pokemon, apps
Analytics	Deploying solutions for asset tracking and management.	Business Intelligence, sales and distribution.
Emergency support	 	Emergency, Disaster recovery services.

Table 1.
Location-based service applications.

provide additional information, and guide the consumer to their desired product. Some method for handover of such applications is needed, without requiring the consumer to re-input product preferences. The GIS area of potential future directions, it is evident that location is merely a starting point for personalisation and context-aware services that use other relevant information when constructing service offers. Moreover, the rapid deployment of alternative wireless technologies, such as WiFi (802.11) is an opportunity and threat for cellular operators, and will likely shape the future development of LBS. Some of the business giants are offering free WiFi service with the advertisement, and even telecom operators are also in this line.

13. Types of location-based services

Pull: In the case of a pull service the user issues a request and access the LBS he/she wants. The requirements or interest of the user is displayed. For, e.g., a tourist roams in a strange city and wants to obtain information about the closest restaurants to his current position.

Push: in the case of a push service, the service provider issues request to the user. For, e.g., location-based advertising, which informs users about the merchandise of their interest located at nearby shops.

Tracking: In a tracking service, the primary thought is that someone (user or service) puts out a request to place the other mobile station (users, vehicles, fleet). Courier tracking, tourist vehicle tracking, With the entire fleet on the central map, the solution led to the best use of the fleet through route optimisation, resource utilisation reports, intelligent trip analytics, online and SMS-based data management. The solution provided has made it efficient and productive also making the vehicles safe for transportation. School Bus Tracking is designed to manage the fleet of buses effortlessly by transforming them into Smart Buses. It guarantees the safety of a minor. Parents stay aware of the whereabouts of children at all times. Parents also kept updated using this service.

14. Opportunities with LBS and VGI technology

Availability of cheaper GPS enabled devices, the combination of GIS and GPS and internet has revolutionised the thinking and the service sector. The GPS provides the location data, and the GIS provides the connection regarding service availability. The Cab service which used the M-Commerce technology was able to provide the economical service as the nearest cab gets a ping for the request and upon the acceptance, the service rendered within a short duration. Likewise, even the user can ascertain the availability of cabby observing the position of the taxi. The beauty of the service was even commuters are pooled en route, thereby providing economic service, saving time and conserving environmental resources. The larger cities are having the issue of parking people prefer to use the service at doorsteps; the eateries are now come up with a solution to supply to their doors the Swiggy, Ubereats and Zomato are helping in this regard. The E-Commerce is also utilising the tracking service for the movement of commodities. The consumers and supplier chains are discovering the solutions by establishing in ideal locations. These service sector work opportunities are engaging a prominent figure of the younger contemporaries in their startups. The travel time is helpful in city environments for estimation and trip plan this provision will help even the workers save time and energy with optimal planning and performance. As pervasive and ubiquitous computing concept becomes more and more popular, accurate positioning plays a vital role in the scenario, embedded GPS in smartphones with assisted-GPS-marked a significant improvement in obtaining quick, precise positioning. While location-based applications usually need precise geographic coordinates. Upon the precise position on the map then LBS is requested. The medicine and diagnostic labs are provided services at doorsteps by the pharmaceutical shops and diagnostic laboratories. Some of the non-critical patients can avail the limited service from the doctors and nurses at their residence without visiting the hospital. These services have saved energy, money, time of patient and caretakers.

15. LBS in real estate

The initiative like the E-Governance, E-Commerce to publish the spatial information of property details of Government/public lands along with land use restrictions for private properties. Thus, by saving time and money spent on legal proceedings. Thus mitigates the property encroachments. The crowdsourcing and the civil policing can fetch the latest information on incidences of encroachments, thus reducing the patrolling cost for the authorities.

Fast actions in cases of reports can be persuaded, thus reducing requisition and demolition costs drastically. Crowdsourcing and publication of spatial information on government properties and land use restrictions can prevent the further land encroachments. The data collected from healthy group volunteers will always be in quantitatively and qualitatively massive comparison with the one organisation and will lead towards big data in the future course. With the GIS the facility there is increased, refinement in the search. The geographical search for buildings can base on the criteria of school transport facility, sports complex, health aspects will narrow down to the required information. With the help of VGI service providers have come up for LBS of regular property visits with detailed reports including photo and video coverage(upon the long absence), collecting mail, water and electric bills and their payments with payment confirmation reports, collecting rents, assisting in insurance plans and claims, legal service regarding the documentation and encumbrances.

16. Conclusion

The roadside assistance from the vehicle companies with web-GIS support becomes a reality. The E-Commerce has captured the market because of the quick supply, competitive rates, easy returns and quicker installation and service. The E-market software has established a link between the suppliers and consumers; the manufacturer also started supplying goods to consumers as the delivery part is taken care of by courier companies. Manufacturers should ensure a continuous supply of materials for supplying the consumer demands. The geographical relation where these transaction is happening is related, this location-based information helps in indicating the overall consumer needs. The geographical information is necessary while estimating the travel time, apart from this E-Commerce; the health sector got an advantage from the VGI. The movement of the emergency vehicle with an allotment to duty to the nearest the vehicle for service has saved many lives. The dedicated sensors can provide the latest information and can be assisted with General Packet Radio Service (GPRS) technology to send the information at regular intervals. The temperature, pressure, wind speed, humidity, rainfall and traffic updates monitored at base stations send the information for recording and processing. Forecasted weather information sent to the entire available user (subscribers) based on the user's location. The artificial intelligence which uses the latest sensor information in the background guides us about the all possible opportunities (upon low fuel indicator in the car it informs about nearest fuel filling station).

The bio-information and geo-information combined to become the gateway to new possibilities to explore. The gadgets becoming more and more personal possessed with all the personal documentation, the necessity of bio- authentication and geographical tracking becomes a necessity. The bio-authentication at specific geographical location reduces crime rates. The accurate GIS data used for delivery of goods by drone is in the experimental stage.

Automated driving with the GPS, GIS and sensors are also gradually under progress with navigation and guidance, driving and safety, with performance. However, more and more information adding to big data the possibility of automation in all fields is replacing the human intervention, thereby reducing human exploitation. With time scale many opportunities are explored simultaneously. The commercial space which is very expensive for business establishment, entrepreneurs using VGI and LBS technologies in their business models established warehouse in remote less expensive areas.

DOI: http://dx.doi.org/10.5772/intechopen.82004

Author details

Nuggehalli Narayanachar Ramaprasad[1*] and Priya Narayanan[2]

1 Department of Geography, School of Earth Science, Central University of Karnataka, Kalaburagi, India

2 National Assessment and Accreditation Council (NAAC), Nagarbhavi, Bangalore, India

*Address all correspondence to: nnrprasad@gmail.com

References

[1] Malage O. Location Based Services. 2016, 25th September 2018. Available from: https://www.tatateleservices.com/downloads/dobigwebinar/resources/brochure/Location_Based_Services.pdf

[2] Simperl E. How to use crowdsourcing effectively: Guidelines and examples. Liber Quarterly. 2015;**25**(1):18-39

[3] Yu L, Gong P. Google Earth as a virtual globe tool for Earth science applications at the global scale: Progress and perspectives. International Journal of Remote Sensing. 2012;**33**(12): 3966-3986

[4] Pringle H. Google Earth shows clandestine worlds. Science. 2010;**329**: 1008-1009

[5] Smith TM, Lakshmanan V. Real-time, rapidly updating severe weather products for virtual globes. Computers & Geosciences. 2011;**37**:3-12

[6] World Commission on Environment Development. In: Brundtland GH, editor. Our Common Future. Oxford: Oxford University Press; 1987

[7] Ehlers M et al. Advancing Digital Earth: Beyond the next generation. International Journal of Digital Earth. 2014;7(1):3-16

[8] Goodchild MF. Citizens as sensors: the world of volunteered geography. Geo Journal. 2007;**69**(4):211-221

[9] Mooney P, Corcoran P. Has OpenStreetMap a role in Digital Earth applications? International Journal of Digital Earth. 2013;7(7):534-553

[10] Stensgaard A et al. Virtual globes and geospatial health: The potential of new tools in the management and control of vector-borne diseases. Geospatial Health. 2009;**3**:127-144

[11] Ghosh S et al. Cyber GIS and crowdsourcing—A new approach in E-Governance. Geospatial World. 2012

[12] Lauriault TP, Mooney P. Crowdsourcing: A Geographic Approach to Public Engagement. Programmable City. 2014. Available from: http://www.nuim.ie/progcity/ [cited Working Paper 6 2 oct 2014]

[13] Richter KF, Winter S. Citizens as database: Conscious ubiquity in data collection. In: Pfoser D et al., editors. Advances in Spatial and Temporal Databases. Lecture Notes in Computer Science. Vol. 6849. Berlin/Heidelberg: Springer; 2011. pp. 445-448

[14] Slocum TA et al. Thematic Cartography and Geovisualization. Prentice Hall; 2008. r. edition

[15] Schnürer R, Sieber R, Çöltekin A. The Next Generation of Atlas User Interfaces: A User Study with Digital Natives Modern Trends in Cartography. Springer International Publishing; 2014. pp. 23-36

[16] Knapp L. A task analysis approach to the visualization of geographic data. In: Nyerges TL et al., editors. Cognitive Aspects of Human-Computer Interaction for Geographic Information Systems. Netherlands: Springer; 1995

[17] Slocum TA et al. Cognitive and usability issues in geovisualization. Cartography and Geographic Information Science. 2001;**28**(1):61-75

[18] Montello DR. Cognitive map-design research in the twentieth century: Theoretical and empirical approaches. Cartography and Geographic Information Science. 2002;**29**(3)

[19] Çöltekin A, Fabrikant SI, Lacayo M. Exploring the efficiency of users' visual analytics strategies based on sequence

analysis of eye movement recordings. International Journal of Geographical Information Science. 2010;**24**(10): 1559-1575

[20] Roth RE. An empirically-derived taxonomy of interaction primitives for interactive cartography and geovisualization. IEEE Transactions on Visualization and Computer Graphics. 2013;**19**(12):2356-2365

[21] Haklay M et al. Crowdsourced geographic information use in government global facility for disaster reduction and recovery (GFDRR). 2014. Available from: http://discovery.ucl.ac.uk/1433169/ [cited 2014 15 October 2014]

[22] Martin C. Barriers to the open government data agenda: Taking a multi-level perspective. Policy and Internet. 2014;**6**(3):217-256

[23] Boyd DS, Foody GM. An overview of recent remote sensing and GIS based research in ecological informatics. Ecological Informatics. 2011;**6**:25-36

[24] Van de Kar E, Bouwman H. The development of location based mobile services. Proceedings of the 4th Edispuut Conference, Amsterdam. 17th October 2001

[25] Searby J. Personalization—An overview of its use and potential. BT Technology Journal. 2003;**21**(1):13-19